# How To
# Repair
# Diesel
# Engines

For my teacher.
Dr. W. P. Baker.
University of Calgary.
Calgary. Alberta. Canada

No. 817
$9.95

# How To Repair
# Diesel
# Engines

## By Paul Dempsey

**TAB BOOKS**
Blue Ridge Summit, Pa. 17214

# Contents

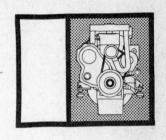

# Preface

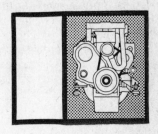

Most mechanics shy away from diesels—very likely because of general feeling that there *has* to be something mysterious and unnatural about an engine that doesn't even use spark plugs. But the diesel is no more complicated than its gasoline-driven counterpart—as this book proves.

Yet, diesel mechanics are scarce and repairs are expensive, even for the ubiquitous power plants in the 3—300 hp range—the kind we find in tractors and other farm implements, boats, air compressors, generators, and a growing number of automobiles. These are the engines that tend to be neglected because there simply aren't enough mechanics with the know-how to cope with them, and with their more imagined than real peculiarities.

Most of the parts of a diesel are very similar to those found in gasoline engines. The crankshaft, pistons, and bearings are merely more massive. The differences are a matter of the unique diesel principle. Fuel is admitted after the compression stroke has been initiated. This means that a very high-pressure fuel pump must be used to overcome cylinder compression and to atomize the fuel oil. High pressures call for high levels of precision, of the kind unknown in gasoline engines. Ignition is by heat of compression, which in turn means heavy bearing loads and low engine speed. Lubrication systems are usually more sophisticated than those found on gasoline engines, and starting systems have a greater capacity to overcome the high cylinder compression. And extreme cold weather starting requires special techniques.

But servicing is not difficult. It merely requires some appreciation of the principles involved and a willingness to perform precision work.

The diesel engine is becoming more and more important as fuel costs increase. No other power plant matches its fuel efficiency and no internal combustion engine is so indiscriminate about the quality of fuel it burns. In addition the diesel is durable and reliable. Highway diesels log a half-million miles and more between major overhauls. These factors tend to compensate for the weight of the engine and its rpm limitations. Consequently, more and more mechanics will be called upon to service these engines. This book is a beginning in that direction

Paul Dempsey

# How To
# Repair
# Diesel
# Engines

# Rudolf Diesel

**1**

Rudolf Diesel was born of German parentage in Paris in 1858. His father was a self-employed leather worker who, by all accounts, managed to provide only a meager income for his wife and three children. Their stay in the City of Light was punctuated by frequent moves from one shabby flat to another. Upon the outbreak of the Franco-Prussian War in 1870, the family became political undesirables and were forced to emigrate to England. Work was almost impossible to find and, in desperation, Rudolf's parents sent the boy to Augsburg to live with an uncle. There he was enrolled in school.

Diesel's natural bent was for mathematics and mechanics. He graduated at the head of his class, and on the basis of his teachers' recommendations and a personal interview by the Bavarian director of education, he received a scholarship to the prestigious Polytechnikum in Munich.

His professor of theoretical engineering was the renowned Carl von Linde, who invented the ammonia refrigeration machine and devised the first practical method of liquefying air. Linde was an authority on thermodynamics and high-compression phenomena. During one of his lectures he remarked that the steam engine had a thermal efficiency of from 6 to 10%; that is, one-tenth or less of the heat energy of its fuel was used to turn the crankshaft—the rest was wasted. Diesel made special note of this fact. In 1879 he asked himself whether heat could not be directly converted into mechanical energy instead of first passing through a working fluid such as steam.

On the final examination at the Polytechnikum, Diesel achieved the highest honors yet attained at the school. Professor Linde arranged a position for the young diploma engineer in Paris, where, in few months, he was promoted to general manager of the city's first ice-making plant. Soon he took charge of distribution of Linde machines over southern Europe.

By the time he was 30 Diesel had married, fathered three children, and was recognized throughout the European scientific community as one of the most gifted engineers of the age. He presented a paper at the Universal Exposition held in Paris in 1889—the only German so honored. When he received the first of several citations of merit from a German university, his acceptance speech was wryly ironic: "I am an iceman..."

The basis of this acclaim was his preeminence in the new technology of refrigeration, his several patents, and a certain indefinable air about the young man that marked him as extraordinary. He had a shy, self-deprecating humor and an absolute passion for factuality. Diesel could be abrupt when faced with incompetence and was described by relatives as "proud." At the same time he was sympathetic to his workers and made friends among them. It was not unusual for Diesel to wear the blue cotton twill which was the symbol of manual labor in the machine trades.

He had been granted several patents for a method of producing clear ice which, since it looked like natural ice, was much in demand by the upper classes. Professor Linde did not approve of such frivolity, and Diesel turned to more serious concerns. He spent several years in Paris, working on an ammonia engine, but in the end was defeated by the corrosive nature of this gas at pressure and high temperatures.

The theoretical basis of this research was a paper published by N. L. S. Carnot in 1824. Carnot set himself to the problem of determining how much work could be accomplished by a heat engine employing repeatable cycles. He conceived the engine drawn in Fig. 1-1. Body 1 is heated; it can be a boiler or other heat exchanger. The piston is at position $C$ in the drawing. As the air is heated it expands in correspondence to Boyles' law. If we assume a frictionless engine, its temperature will not rise. Instead, expansion will take place, driving the piston to $D$. Then $A$ is removed, and the piston continues to lift to $E$. At this point the temperature of the air falls until it exactly matches cold surface 2 (which can be a radiator or cooling tank). The air column is now placed in contact with 2, and the piston falls because the air is compressed. Note, however, the temperature of the air does not change. At $B$ cold body 2 is removed, and the piston falls to $A$. During this phase the air gains temperature until it is equal to 2. The piston climbs back into the cylinder.

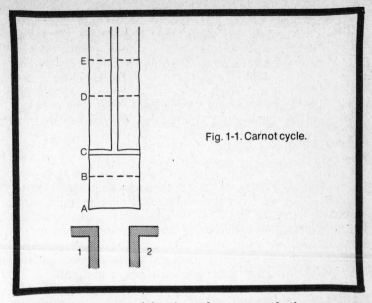

Fig. 1-1. Carnot cycle.

The temperature of the air, and consequently the pressure, is higher during expansion than during compression. Since the pressure is greater during expansion, the power produced by the expansion is greater than that consumed by the compression. The net result is a power output that is available for driving other machinery.

Of course this is an "ideal" cycle. It does not take into account mechanical friction nor transfer of heat from the air to the piston and cylinder walls. The infinitesimal difference of heat between 1 and 2 is sufficient to establish a gradient and drive the engine. It would be completely efficient.

In 1892 and 1893 Diesel obtained patent specifications from the German government covering his concept for a new type of *verbrennungskraftmachinen*, or heat engine. The next step was to build one. At the insistence of his wife, he published his ideas in a pamphlet and was able to interest the leading Augsburg engine builder in the idea. A few weeks later the giant Krupp concern opened negotiations. With typical internationalism he signed another contract with the Sulzer Brothers of Switzerland.

The engine envisioned in the pamphlet and protected by the patent specifications had these characteristics:

- *Compression of air prior to fuel delivery.* The compression was to be *adiabatic*; that is, no heat would be lost to the piston crown or cylinder head during this process.

- *Metered delivery of fuel so compression pressures would not be raised by combustion temperatures.* The engine would operate on a *constant—pressure cycle;* expanding gases would keep precisely in step with the falling piston. This, of course, is a salient characteristic of Carnot's ideal gas cycle, and stands in contrast to the Otto cycle, in which combustion pressures rise so quickly upon spark ignition that we describe it as a *constant-volume* engine.
- *Adiabatic expansion.*
- *Instantaneous exhaust at constant volume.*

It is obvious that Diesel did not expect a working engine to attain these specifications. Adiabatic compression and exhaust phases are, by definition, impossible unless the engine metal is at combustion temperature. Likewise, fuel metering cannot be so precise as to limit combustion pressures to compression levels. Nor can a cylinder be vented instantaneously. But these specifications are significant in that they demonstrate an approach to invention. The rationale of the diesel engine was to save fuel by as close an approximation to the Carnot cycle as materials would allow. The steam, or *Rankine cycle*, engine was abysmal in this regard; and the *Otto* 4-stroke-cycle spark or hot-tube-ignition engine was only marginally better.

This approach, from the mathematically ideal to materially practical, is exactly the reverse of that favored by inventors of the Edison, Westinghouse, and Kettering school. These men, as Thomas A. Edison put to Diesel when Diesel visited America in 1912, worked *inductively*, from the existing technology, and not *deductively*, from some ideal or model. Diesel felt that such procedure was at best haphazard, even though the results of Edison and other inventors of the inductive school were obviously among the most important. Diesel believed that productivity should be measured by some absolute scientific standard.

The first Diesel engine was a single-cylinder 4-cycle design, operated by gasoline vapor. The vapor was sprayed into the cylinder near top dead center by means of an air compressor. The engine was in operation in July, 1893. However it was discovered that a misreading of the blueprints had caused an increase in the size of the chamber. This was corrected with a new piston, and the engine was connected to a pressure gage. The gage showed approximately 80 atmospheres before it shattered, spraying the room with brass and glass fragments. The best output of what Diesel called his "black mistress" was slightly more than 2 hp—not enough power to overcome friction and compression losses. Consequently, the engine was redesigned.

The second model was tested at the end of 1894. It featured a variable-displacement fuel pump to match engine speed with load. In February of the next year, the mechanic Linder noted a remarkable development. The engine had been sputtering along driven by a belt from the shop power plant, but Linder noticed that the driving side of the belt was slack, indicating that the engine was putting power into the system. For the first time the Diesel engine ran on its own.

Careful tests—and Diesel was nothing if not careful and methodical—showed that combustion was irregular. The next few months were devoted to redesign of the nozzle and delivery system. This did not help, and in what may have been a fit of desperation, Diesel called upon Robert Bosch for an ignition magneto. Bosch personally fitted one of his low-tension devices to the engine, but without much effect on the combustion problem. Progress came about by varying the amount of air injected with the fuel, which, at this time, was limited to kerosene or gasoline.

A third engine was built with a smaller stroke/bore ratio and fitted with two injectors. One delivered liquid fuel, the other a mixture of fuel and air. This was quite successful, producing 25 hp at 200 rpm. It was several times as efficient as the first model. Further modifications of the injector, piston, and lubrication system ensued, and the engine was deemed ready for series production at the end of 1896.

Diesel turned his attention to his family, music, and photography. Money began to pour in from the patent licensees and newly organized consortiums wanting to build engines in France, England, and Russia. The American brewer Adolphus Busch purchased the first commercial engine, similar to the one on display at the Budweiser plant in St. Louis today. He acquired the American patent rights for one million marks which at the current exchange rate amounted to a quarter of a million dollars—more than Diesel had hoped for.

The next stage of development centered around various fuels. Diesel was already something of an expert on petroleum, having researched the subject thoroughly in Paris in an attempt to refine it by extreme cold. It soon became apparent that the engine could be adapted to run on almost any hydorcarbon from gasoline to peanut oil. Scottish and French engines routinely ran on shale oil, while those sold to the Nobel combine in Russia operated well on refinery tailings. In a search for the ultimate fuel, Diesel attempted to utilize coal dust. As dangerous as this fuel is in storage, he was able to use it in a test engine.

These experiments were cut short by production problems. All of the licensees did not have the same success with the engine. In at least one instance, a whole production run had to be recalled. The difficulty was further complicated by a shortage of trained technicians. A small malfunction could keep the engine idle for weeks, until the customer lost patience and sent it back to the factory. With these embarrassments came the question of whether the engine had been oversold. Some believed that it needed much more development before being put on the market. Diesel was confident that his creation was practical—if built and serviced to specifications. But he encouraged future development by inserting a clause in the contracts which called for pooled research; the licensees were to share the results of their research on Diesel engines.

Diesel's success was marred in two ways. For one thing he suffered exhausting patent suits. The Diesel engine was not the first to employ the principle of compression ignition; Akroyd Stuart had patented a superficially similar design in 1890. Also, Diesel had a weakness for speculative investments. This weakness, along with a tendency to maintain a high level of personal consumption, cost Rudolf Diesel millions. His American biographers, W. Robert Nitske and Charles Morrow Wilson, estimate that the mansion in Munich cost a million marks to construct at the turn of the century.

The inventor eventually found himself in the uncomfortable position of living on his capital. His problem was analogous to that of an author who is praised by the critics but who cannot seem to sell his books. Diesel engines were making headway in stationary and marine applications. But they were expensive to build and required special service techniques. True mass production was out of the question At the same time, the inventor had become an international celebrity, acclaimed on three continents.

Diesel returned to work. After mulling a series of projects, some of them decidedly futuristic, he settled on an automobile engine. Two such engines were built. The smaller, 5 hp model was put into production, but sales were disappointing. The engine is, by nature of its compression ratio heavy, and, in the smaller sizes, difficult to start. (The latter phenomenon is due to the unfavorable surface/volume ratio of the chamber as piston size is reduced. Heat generated by compression tends to bleed off into the surrounding metal.) A further complication was the need for compressed air to deliver the fuel into the chamber. Add to these problems precision machine work, and the diesel auto engine seemed impractical. Mercedes-Benz offered a diesel-powered passenger car in 1936. It was followed by the Austin taxi (remembered with mixed feelings by travelers to postwar London), by the Land Rover, and more recently, by the Peugeot. However desirable diesel cars are from the point of view of fuel economy and longevity, they are still not competitive with gasoline-powered cars.

Diesel worked for several months on a locomotive engine which was built by the Sulzer Brothers in Switzerland. First tests were disappointing, but by 1914 the Prussian and Saxon State Railways had a diesel in everyday service. Of course, most of the world's locomotives are diesel powered today.

Maritime applications came as early as 1902. Nobel converted some of his tanker fleet to diesel power, and by 1905 the French navy was relying upon these new engines for their submarines. Seven years later, almost 400 boats and ships were propelled solely or in part by compression engines. The chief attraction was the space saved, which increased the cargo capacity or range.

In his frequent lectures Diesel summed up the advantages of his invention. The first was efficiency, which was beneficial to the owner and, by extension, to all of society. In immediate terms efficiency meant cost savings. In the long run it meant conserving world resources. Another advantage was that compression engines could be built on any scale from the 2400 hp Italian Tosi of 1912 to the fractional horsepower Compared to steam engines the diesel is compact. Also, the compression engine is clean. Rudolf Diesel was very much concerned with the question of air pollution, and mentioned it often.

But the quintessential characteristic, and the one which may explain his devotion to his "black mistress," was her quality. Diesel admitted that the engines were expensive, but, his goal was to build the best, not the cheapest.

During this period Diesel turned his attention to what his contemporaries called "the social question." He had been poor and had seen the effects of industrialization firsthand in France, England, and Germany. Obviously machines were not freeing men, or at least not the masses of men and women who had to regulate their lives by the factory system. This paradox of greater output of goods and intensified physical and spiritual poverty had been seized on by Karl Marx as the key "contradiction" of the capitalistic system. Diesel instinctively distrusted Marx because he distrusted the violence which was inplicit in "scientific socialism." Nor could he take seriously a theory of history which purported to be based on absolute principles of mathematical integrity.

He published his thoughts on the matter under the title *Solidarismus* in 1903. The book was not taken seriously by either the public or politicians. The basic concept was that nations were more alike than different. The divisions which characterize modern society are artificial to the extent that they have any other than an economic rationale. To find solidarity the mass of humanity must become part owners in the sources of production. His formula was for every worker to save a penny a day. Eventually these pennies would add up to shares or part shares in business enterprises. Redistributed wealth and, more important, the sense of control of one's destiny would be achieved without violence or rancor through the effects of the accumulated capital of the workers.

Diesel wrote another book which was better received. Entitled *Die Enstehung des Dieselmotors*, it recounted the history of his invention. It was published in the last year of his life.

For years he had suffered migraine headaches, and in his last decade, he developed gout, which at the end forced him to wear a special oversized slipper. Combined with this was a feeling of fatigue, a sense that his work was both done and undone, and that there was no one to continue. Neither of his two sons showed any interest in the engine, and he himself seemed to have lost the iron concentration of earlier years when he thought nothing of a 20-hour working day. It is probable that technicians in the various plants knew more about the current state of diesel development than he did.

And the bills mounted. A consultant's position, one that he would have coveted in his youth, could only forestall the inevitable; a certain level of indebtedness makes a salary superfluous. Whether he was serious in his acceptance of the English-offered consultant position is unknown. He left his wife in Frankfort in apparent good spirits and gave her a present. It was an overnight valise, and she was instructed not to open it for a week. When she did she found it contained 20,000 marks. This was, it is believed, the last of his liquid reserves. At Antwerp he boarded the ferry to Harwick in the company of three friends. They had a convivial supper on board and retired to their staterooms. The next morning Rudolf Diesel could not be found. One of the crew discovered his coat, neatly folded under a deck rail. The captain stopped the ship's progress, but there was no sign of the body. A few days later a pilot boat sighted a body floating in the channel, removed a coin purse and spectacle case from the pockets, and set the corpse adrift. The action was not unusual or callous; seamen had, and still do have, a horror of retrieving bodies from the sea. These items were considered by the family to be positive identification. They accepted the death as suicide, although the English newspapers suggested foul play at the hands of foreign agents who did not want Diesel's engines in British submarines.

# Diesel Basics 2

If we discount the pump and injectors, a diesel engine looks like a very sturdy gasoline engine. Both use much the same parts (see Fig. 2-1), and they operate on similar though not identical cycles. Diesel engines can be made to operate on gasoline (one of the first to show this capability was the GM 2-cycle), and gasoline engines have been converted to diesel operation. The British Rover engine, a V-8 designed for armored vehicles, can be modified in the field to convert it from gasoline to heavy fuel.

The major difference between the diesel and the gasoline engine is the method of ignition. A diesel has no ignition system as such. The air/fuel mixture is ignited by heat generated during the compression stroke. Heat can be thought of in terms of a molecular motion. The hotter a substance is, the faster its molecules vibrate. When a gas is compressed part of the energy of compression is transferred to the molecules as velocity increases. When a gas expands it loses heat.

Diesel engines compress air to a volume of $1/17$ to $1/23$ its original volume. In the process the temperature of the air increases to a minimum of 750°F, enough to produce ignition. Gasoline engines are limited by fuel type and mechanical strength to compression ratios of no more than 11:1 for high-performance engines, and about 7:1 for industrial types—far below the ignition point.

Ignition in a gasoline motor is by a high-voltage spark which arcs between the electrodes of the spark plug. The spark is timed to occur at or a few degrees before the piston reaches the upper limit of travel.

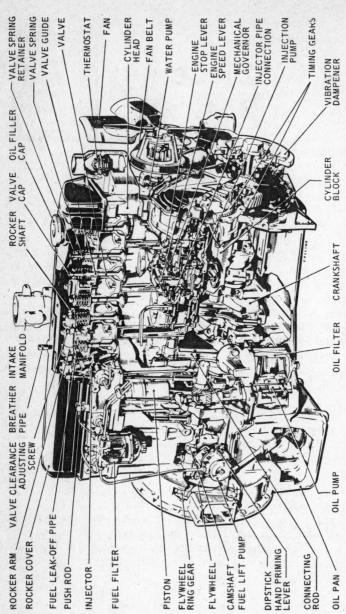

ROCKER ARM — VALVE CLEARANCE — BREATHER — INTAKE — ROCKER — VALVE — OIL FILLER — VALVE SPRING RETAINER

ROCKER COVER — ADJUSTING SCREW — PIPE — MANIFOLD — SHAFT — CAP — CAP — VALVE SPRING

VALVE GUIDE

FUEL LEAK-OFF PIPE

VALVE

PUSH ROD

THERMOSTAT

INJECTOR

FAN

FUEL FILTER

CYLINDER HEAD

FAN BELT

WATER PUMP

PISTON

ENGINE STOP LEVER

FLYWHEEL RING GEAR

ENGINE SPEED LEVER

FLYWHEEL

MECHANICAL GOVERNOR

CAMSHAFT

INJECTOR PIPE CONNECTION

FUEL LIFT PUMP

INJECTION PUMP

DIPSTICK

VIBRATION DAMPENER

HAND PRIMING LEVER

TIMING GEARS

CONNECTING ROD

CYLINDER BLOCK

OIL PAN

OIL PUMP

OIL FILTER

CRANKSHAFT

Fig. 2-1. Diesel nomenclature. (Courtesy Industrial Engine and Turbine Operations, Ford Motor Co.)

22

Ignition in a diesel engine is not timed, but is a function of air density, engine temperature, and fuel blends. Consequently, the diesel compresses only air. Fuel is introduced at the end of the compression stroke, shortly before the piston reaches top dead center. The introduction of the fuel synchronizes combustion with piston movement.

The requirement that fuel be forced into the cylinder against compression adds much to the cost and complexity of the diesel. The fuel system consists of a low-pressure or *transfer pump*, to move the fuel from the tank; a high-pressure pump; and one injector per cylinder, which functions as a fuel entry point and as a check valve against the loss of engine pressure into the fuel lines. These parts may be separate entities or in combination with each other. Great precision is required for the parts which meter fuel.

In theory, a diesel engine requires 10,000 volumes of air per volume of fuel oil. As a practical matter the air figure is increased by one quarter to reduce smoke. If we consider engine as an air pump we can see that the fuel pump section must have 1/12,500 the capacity of the engine cylinders. The scale effect is formidable and is made more so by the fact that fuel must be injected under pressure. Not only does the fuel have to be forced into the cylinder, it must be atomized into a fine spray and be able to penetrate deeply into the chamber. Delivery pressures vary widely. An average range is between 2000 and 6000 psi, but with some engines, delivery pressure reaches 20,000 psi.

## FUEL METERING

Fuel metering parts must be manufactured with the greatest precision. Ordinary production line techniques are inadequate. Critical assemblies such as high-pressure pumps and barrels are lapped individually to extremely close tolerances. In practical terms this means that a new pump cannot be assembled if the plungers have been in direct sunlight and the barrels in the shade!

The fuel must be filtered to remove solids, which would quickly destroy the metering parts. Several filters are connected in series between the pump and the tank. Most small engines incorporate a sediment bowl at the primary filtering stage to catch any water which might be in the system. Water is such a critical contaminant that some manufacturers of vehicle and boat engines suggest that they not be refueled during rain.

## WEIGHT

High compression ratios are responsible for another characteristic of the engine—its weight. The forces generated prior to and during combustion can only be contained by making the pistons, connecting rods, and block heavier than they would be for an equivalent gasoline engine. The first diesel weighed 450 lb per developed horsepower. This figure was reduced in production engines, and today the weight-to-horsepower ratio for vehicle plants is in the neighborhood of ¾:1. This remarkable improvement has come about from innovations in many areas, the most significant of which have been the use of *solid injectors* (rather than air injectors, which required multistage air compressors); improved combustion, achieved through chamber and nozzle design; light-alloy pistons (which mean better heat conductivity and lighter reciprocating masses); and *turbocharging*.

Turbocharging adds only a few pounds to the engine, and can boost power by as much as 50%. For example, the Greenwich Marine engine delivers 136 hp in its normally aspirated form; with turbocharging, this jumps to 190 hp. However, the diesel has not caught up with the spark ignition engine in this regard. Perhaps it never will. Both engines draw from the same broad collection of engineering and theoretical knowledge. Improvements in one are, in many cases, adapted to the other. This was true of supercharging, swish-type combustion chambers, and tuned exhaust porting.

## THERMAL EFFICIENCY

High compression ratios are not without their virtues. High compression ratios—or more exactly, large ratios of expansion—mean improved *thermal efficiency*. Thermal efficiency is the measure of how effectively the heat in the fuel is converted into work. Well designed spark ignition engines have thermal efficiencies of 30% or so. About one-third of the heat latent in the gasoline is recovered as work. The rest is lost through the cooling system or exhaust or dissipated as friction. Diesel engines have thermal efficiencies in the neighborhood of 40%, making them the most efficient of all engine types.

Thermal efficiency has a direct bearing on fuel economy. Diesels are inherently economical, and become more so when operated at part load. Diesel fuel remains cheaper than gasoline (although this could change if the demand for the fuel increases, since there is only so much distillate per barrel of crude) and is heavier. The weight difference is significant since the engine extracts thermal energy from the fuel by the pound, while the diesel operator buys fuel by the gallon.

Diesel fuel is safe (usually has a flash point of 150°) and storable. Gasoline has an open shelf life of about 6 months and, depending upon the blend, may have a flash point as low as 45°F. But this is not to say that fuel oil should be treated carelessly. The fumes which escape from tank vents, spillage, and during refueling are explosive and can be ignited by spark or open flame.

It is generally agreed that diesel engines are longer-lived than their gasoline counterparts. Hard comparisons are difficult to find—perhaps because so many diesel manufacturers also produce spark ignition engines—but the testimony of mechanics and diesel operators supports this contention. Peugeot, the French car maker which has recently introduced a 4-cylinder diesel passenger car, claims 150,000 miles between overhauls in normal service. General Motors warrant its intracity bus engines for 100,000 mi.

*Reliability*, or the tendency of the engine to function without sudden and unpredictable failure, is enhanced by the absence of spark plugs and ignition contact points. These parts pose problems at between 5,000 and 10,000 miles in highway service. Nor does the diesel have a carburetor. Diesel accessories—fuel, water, and oil pumps; alternators; and radiators—tend to be heavier and better quality than those fitted to equivalent spark ignition engines.

*Durability*, or the resistance to wear and slow degradation of performance, is aided by the massive construction of the block, connecting rods, crankshaft, and other moving parts. It may be argued that these parts need to be strong because of the stresses involved, but the sum total of the effect is rigidity and predictable wear patterns. But the major durability factor is that diesels are understressed. These engines work with an excess of air to retard exhaust smoke. Consequently, power is down from what it theoretically could be. Also, diesels are operated at moderate speeds. Perhaps the highest diesel revving engine on record is the diminutive Peugeot 204, which turns at 5000 rpm. Most diesels are governed to less than 4000 rpm, and spend their working lives at considerably more leisurely speeds.

Another power-limiting factor for diesels is the size of the intake valves. High compression ratios mean that the combustion chamber space is severely restricted. Really large-diameter valves cannot be used (although multiple valves can increase the port area by as much as 40% over single intakes and exhausts), and the lift is restricted by the proximity of the piston.

## THE NATURE OF DIESEL COMBUSTION

Diesel engines are clean. The diesel's smoke and characteristic odor hardly seem to square with that statement, but nevertheless it is true in terms of hydrocarbons, carbon monoxide, and nitrous oxide emissions. These three pollutants are considered the most toxic to animal and plant life, and are the only ones currently regulated by law. The Environmental Protection Agency has tested a number of diesel passenger cars and found them to be well under the 1976 standards for hydrocarbon and carbon monoxide. Emissions made up of oxides of nitrogen are a bit over the standard, but can be curtailed by recirculating some of the exhaust gases to the chamber. The effect is to quench the flame, reducing temperatures. Interestingly, fuel economy and driveability do not seem to be affected. The only bad side effect is increased smoke output.

Diesels are clean for several reasons. Excess air is available for full combustion and to prevent the formation of carbon monoxide. Hydrocarbon emissions are down because most of the combustion activity takes place over the piston or, initially, in a carefully shaped alcove leading to the main chamber. The fuel spray is kept clear of remote reaches of the cylinder which would quench the flame. Most hydrocarbon emissions from gasoline engines come about because combustion is snuffed out by the cold cylinder walls.

Even so, diesels smell and smoke. The smell is something that takes some getting used to. (People said the same thing about gasoline engines 70 years ago and we would say the same about horses today.) But the smoke problem is not something one becomes inured to. It is especially severe in out of tune engines, and is present in all of them under part-throttle loads and acceleration.

Chemical additives are some use in limiting smoke. Sullivan Chemical Company has had good success with their Sul-Kem No. 40 fuel additive. Customers testify that smoke is almost eliminated with this product. Apollo Chemical's Dieselex CC-2 has been observed to drop Shell Bacharach smoke number by 2 to 5 units.

## DIESEL VERSATILITY

Another characteristic of diesels is the way in which these engines can be scaled to the application. Some of the Hartz engines discussed in this book develop only 3 hp. Engines in the 20,000 hp class are by no means unusual for stationary and marine applications. One built by the H. C. Oersted Central Electricity Works, in Copenhagen, is typical. This monster has 8 double-acting cylinders, with 33 in. bores and a 59 in. stroke, and is rated at 22,500 hp in continuous duty. No other engine type—whether steam, Stirling, or spark ignition—has shown such versatility. Nor is the diesel locked into the piston concept; engineers at Rolls-Royce built a successful Wankel diesel.

Most diesels are built in a fashion that could almost be called modular. To increase the size of the engine, the manufacturer merely adds cylinders. So long as the center-to-center distance of the bores is not changed this approach can save money and tremendously simplify parts inventories. Thus, the same engine family may have one, two, three, four, five, six, and eight cylinders.

Odd numbers are of little consequence since these engines are sturdy enough to dampen out-of-balance forces and since some vibration is acceptable in most applications. The problem becomes sticky in passenger cars. When Mercedes-Benz decided to enlarge the 4-cylinder 240D, the firm's engineers opted for a 5-cylinder version. This meant a careful redesign of the harmonic balancer, flywheel, and drive line components.

## Operation

Nikolaus August Otto (1832—1891) is generally credited with the development, in 1876, of the engine cycle which bears his name. The Otto working cycle consists of four events which are repeated in sequence as long as the engine runs. During the *intake* cycle air and fuel are drawn into the engine by the downward motion of the piston. This mixture is compressed as the piston moves toward top dead center during the *compression* cycle. Just before the piston reaches the limit of its stroke, the mixture is ignited. Gases given off by combustion drive the piston down on the *expansion*, or *power*, cycle. The spent gases are driven out of the cylinder during the *exhaust* cycle, and the cylinder is ready to be charged with a fresh air/fuel mixture.

These four events can be accomplished in four strokes of the piston (from bottom dead center, or BDC, or to top dead center, or TDC), or can be telescoped into two. Four-cycle engines have one stroke per cycle; providing one expansion stroke every second revolution, or 720° of crankshaft rotation. Two-cycle engines have an expansion stroke during each revolution of the crank. The intake and exhaust cycles are superimposed upon each other.

All spark ignition (SI) engines, including the Wankel rotary, operate upon the Otto working cycle. Dr. Diesel had intended a sharp distinction between his cycle, developed out of an attempt to realize Carnot's principle, and the Otto cycle. One difference is that fuel is injected late into the compression ignition (CI) engine, after the air charge is compressed.

27

Another is that the CI engine draws a constant volume of air (subject only to changes in volumetric efficiency as rpm changes) regardless of rotational speed. At idle a typical diesel has a great surplus of air and "inhales" 100 lb or so per pound fuel. At wide open throttle the ratio drops to 19—20 lb of air per pound of fuel oil. In contrast, the SI engine meters both fuel and air at the carburetor throttle plate. The relative quantities in the mixture change due to air velocity, but the amount of air and fuel vapor going into the engine is proportional to speed and load.

Combustion in the SI engine progresses quickly as the flame front expands through the mixture from the spark plug arc. Pressures rise more rapidly than the piston falls, and for practical purposes the pressure rise is so abrupt that the piston may be considered to be stopped. Hence, the name *constant-volume* engine.

In an attempt to obtain smoother combustion, Diesel specified in his proposal of 1893 that expansion would be at

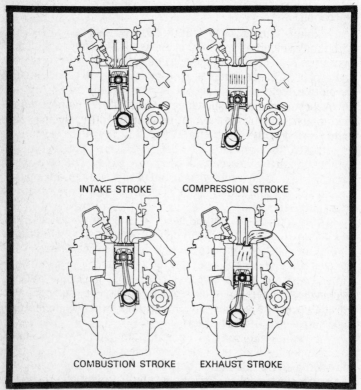

INTAKE STROKE    COMPRESSION STROKE

COMBUSTION STROKE    EXHAUST STROKE

Fig. 2-2. Four-cycle operation. (Courtesy Marine Engine Div., Chrysler Corp.)

constant pressure. That is, fuel would be metered in such a precise and progressive manner that combustion pressures would not rise above compression pressure. In other words, the fuel input would be regulated to keep step with the falling piston. In actual engines this idea was approached, but it required that speeds be kept low. As soon as Dr. Diesel stepped out of the way, his colleagues abandoned the idea as impractical, and controlled fuel input pragmatically, by the power produced. However, diesel engines do not have the sudden pressure rises associated with SI engines and are sometimes with a little exaggeration known as *con stant-pressure* engines.

A 4-cycle diesel engine operates as shown in Fig. 2-2. During the intake stroke the piston draws air into the chamber from the header. The intake valve closes and the air is compressed between the piston crown and the combustion chamber roof. Fuel is injected just before TDC. The fuel ignites because of the increased temperature of the compressed air, and the piston is driven back down the bore on the expansion stroke. Near BDC, the exhaust valve opens and the cylinder blows down until exhaust pressure approaches atmospheric. The valve remains open on the exhaust stroke to insure full evacuation. Valves and injectors are driven at half engine speed to produce one expansion stroke every second revolution.

A 2-cycle engine is shown next (Fig. 2-3). This particular engine (a schematic representation of Detroit Diesel Series 53) employs a Roots blower for unidirectional scavenging and charging. (The term *scavenging* refers to the process of removing exhaust residues from the cylinder in preparation for admitting fuel or *charging*.)

From the left of the drawing, the process is as follows: During the scavenging phase the exhaust valve is open; pressurized air enters through a series of ports milled into the side of the liner and forces the spent gases out around the open valve. The valve closes and the piston compresses the air charge. Near TDC the injector sprays fuel into the chamber and the piston is forced down. But the expansion stroke is cut short by the opening of the exhaust valve just prior to uncovering the air inlet ports. Larger engines sometimes employ a piston-type scavenger pump, although this configuration is considered obsolete in most quarters.

## SCAVENGING SYSTEMS

While the General Motors system is perhaps the most familiar (these engines have been in volume production since

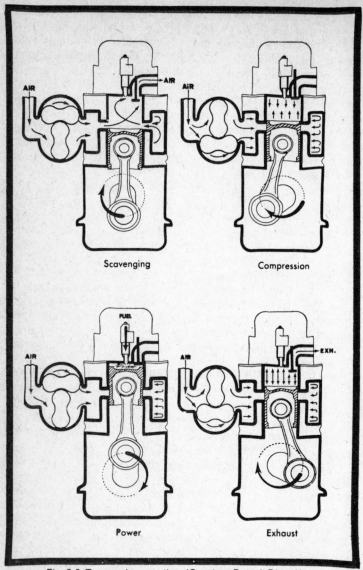

Fig. 2-3. Two-cycle operation. (Courtesy Detroit Diesel.)

1938 and have a scavenging efficiency of 80%), one should realize that there are other methods of controlling scavenging and exhaust gases. One of the most unique was an English design patterned and working in 1928. No valves were used. The exhaust was scavenged by means of pressure waves from a tuned exhaust pipe which were timed to piston movement.

More practical engines include the Nordberg design, which employs automatic valves in the air header to prevent backflow by the exhaust. These valves close when exhaust pressure is higher than scavenging air pressure. A number of rotary-drum and disc-valve arrangements have been used with varying success. The Hamilton engine features a rotary exhaust valve. Scavenging is controlled by a piston port. The rotary valve opens for blowdown and closes before too much fresh air is lost. A number of engines employ air inlet valves and exhaust ports in exactly the reverse of the GM design.

Some 2-cycles dispense with the pump altogether and use the back side of the piston to pressurize the crankcase. This pressure is transferred to the chamber by means of ports opening to the bore. Almost all examples of this type employ exhaust ports which are uncovered by the piston on its downward movement. These ports must, perforce, be placed above the inlet ports from the crankcase.

Most exhaust gases blow down as soon as the piston opens the exhaust ports. However, there is still some residual gas in the cylinder, even though it may be at less than atmospheric pressure due to the inertia of the gas column which has left through the exhaust port. Unlike engines with mechanically coupled or exhaust-driven blowers, crankcase-scavenged engines are "short winded." The only scavenging and charging air available is that generated by the back of the piston. Consequently, cylinder chamber scavenging is quite critical, since there is little surplus for blowthrough.

Two methods are commonly used to clear the cylinders of combustion residues. The earliest was to employ *deflector pistons*. These pistons were domed, with unequal sides. The steep flank was toward the inlet port so the scavenging air would be deflected upward away from the gaping exhaust port. The disadvantages of this arrangement were several. In the first place it was inefficient in terms of air loss and cylinder scrubbing. Secondly, the piston was asymmetrical. Unless the designers were very careful, the piston would expand unevenly. This tendency was abetted by the temperature differential across the deflector. The steep side was cooled by the air charge while the shallow half had almost constant contact with the exhaust gases.

*Loop scavenging*, based on the Schnurle patents in Germany, offers a way out. The piston is flat or slightly dished. Several air inlet ports are milled into the liner and angled so the air rebounds from the top of the chamber. Velocities are very high, and the air takes on a definite swirl, driving the exhaust gases out before it. The only fly in the ointment is that loop- or swirl-scavenged engines are quite temperamental during the design stage. A great deal of cut-and-try experimentation is needed before one can persuade a loop-scavenged engine to run properly. The area, angle, position, and shape of the inlet ports are extremely critical, as is the piston-bore/stroke ratio. Manufacturing to the tolerances required can give rise to unexpected difficulties. In general, greater power output and the promise of better scavenging is worth the effort, and many late-production 2-cycles are built on this principle.

# Fuels and Combustion 3

Petroleum consists of thousands of compounds in various proportions. The great majority of these compounds consist of hydrogen and carbon in various combinations. Oxygen, nitrogen, and sulfur are present either in their elemental forms or in combination. The Baku oil fields produce crude with 87% carbon, 12% hydrogen, and only 1% sulfur. In constrast, California crude is 82% carbon, 10% hydrogen, and 8% sulfur. West Texas production is 85% carbon, 11% hydrogen, and 4% sulfur.

The vast array of hydrocarbons in crude is classified by the number of carbon atoms per molecule and by the formation of the molecule. Gases are, of course, the lightest, with between 1 and 4 carbon atoms. Next comes gasoline, which may have as many as 10 carbon atoms per molecule, although 8 is a more typical figure. Light fuel and heating oils are more complex and have as many as 50 carbon atoms in various configurations. Carbon atoms link to hydrogen in a ring or chain formation, and may have a full complement of hydrogen or may be missing one or more. When the complement is full the molecule is described as *saturated*; when less than complete we say the molecule is *unsaturated*.

Unsaturated molecules are highly reactive—they enter into chemical activity readily since a hydrogen atom does not need to be displaced. These unsaturated hydrocarbons are called *aromatics* if the carbon bond is in the form of a ring. If the bond is a chain the molecules are known as *olefins*. The more stable saturated molecules form the basis of a convenient categorization of crude petroleum. Saturated chain hydrocarbons are known generically as *paraffins*. Pennsylvania and other crudes from the eastern United States

33

are paraffin-based oils since these saturated chain hydrocarbons predominate. Saturated ring molecules are naphthenes. California crude has a high percentage of these and is therefore known as *naphthenic*, or *asphalt-based*. As could be expected, oil from the central U.S. combines both of these compounds and is therefore described as *mixed base*.

As a point of interest, you can distinguish between paraffin-and asphalt-based oils by allowing a few drops to be absorbed on white paper. The paraffin-based oil will not stain the paper, while the asphalt-based oil leaves a brown smudge.

## DIESEL FUEL

Diesel fuel falls into three categories: crude, distillates, and residuals. Because of the CI engine's indiscriminate appetite for hydrocarbons, many of these engines can run on crude after it is centrifuged to remove sand and water. It is not uncommon to find oil rig engines feeding directly from the wellhead. The most strenuous objection to this procedure is economic: Crude is worth too much in potential profit to be consumed in this manner. Another objection concerns the high sulfur content of most crudes. Sulfur impedes upper-cylinder and injection system lubrication and forms compounds which attack the exhaust valves and piping. Sulfur is also a contaminant which, alone or combined, can damage living organisms.

Distillates or, as they are called in the industry, *gas oils*, represent that fraction which is left after gasoline has been boiled off. At the top of the range, distillates include light oils which are nearly identical to kerosene; the low end may consist of heavy saturated compounds with a boiling temperature of 650°F. Almost all diesel fuel is a blend of distillates.

| | No. 1-D | No. 2-D |
|---|---|---|
| 1. Flash pt (°Fmin) | 100 | 125 |
| 2. Cetane No. (min) | 40 | 40 |
| 3. Viscosity at 100°F, | | |
|    centistokes min: | 1.4 | 2.0 |
|    max: | 2.5 | 4.3 |
| 4. Water and sediment | | |
|    (% by volume, max) | Trace | 0.10 |
| 5. Sulfur (% max) | 0.5 | 0.5 |
| 6. Carbon residue (%) | 0.15 | 0.35 |
| 7. Ash (% by wt, max) | 0.01 | 0.02 |
| 8. Distillation (°F) | | |
|    90% pt, max: | 550 | 640 |
|    min: | — | 540 |

Fig. 3-1. ASTM classification of light diesel fuels.

Residuals are what is left at the bottom of the fractionating tower. Although once limited to asphalt, residuals can be converted into lube oils, waxes, and heavy fuel oils. For years engineers have attempted to modify diesel engines to burn residuals. Initially they met with some success, but they have been frustrated by the efficiency of the refineries. Residuals become heavier and more inert as more and more compounds are taken from them.

## Diesel Fuel Specifications

The American Society of Testing Materials (ASTM) has developed standards for diesel fuel oil. These standards are reprinted in Fig. 3-1.

The *flash point* has nothing to do with engine operation. It is the minimum temperature at which oil will release combustible vapors. When ignited these vapors flare and die. Under normal circumstances the oil remains a liquid and does not burn. The significance of flash point is to conform with fire department and insurance regulations as pertains to fuel storage.

The most critical parameter of diesel oil is its *cetane number*. Diesel knock can be minimized by reducing the ignition lag or the time between injection and the beginning of combustion. Cetane is a chain paraffin which gives rapid ignition. It has been assigned a number of 100. Alpha-methyl-naphthalene represents the bottom of the scale, with a cetane number of zero. These two hydrocarbons are mixed and matched against an unknown fuel in a standard research engine. Injection is timed to begin at 13° before TDC, top center, and the compression ratio is varied until ignition occurs at TDC. The percentage of cetane which requires the same compression ratio for TDC ignition is the cetane number of the fuel.

*Viscosity* is an important fuel characteristic, especially in subzero weather. The viscosity of a fuel is its "pourabilty." It is determined with the aid of one of two types of Saybolt viscometers. The number corresponds to the time in seconds for 60 cu cm of fuel to flow through the orifice of the viscometer at standard temperature and atmospheric pressure. The higher the number, the greater the viscosity and the greater the reluctance of the fuel to flow. The pour point represents the nether limit of the viscosity scale. It is the temperature at which the oil solidifies.

Water and solid impurities must be kept to a minimum to keep the filters and screens open and to prevent wear on the injectors, pumps, and other vital parts. Unfortunately, diesel fuel contains more solids than the lighter, more aromatic fuels. The ASTM allows a greater sulfur content for the larger, slow-turning engines than for the smaller types.

Carbon residue is determined by distillation of a fuel sample under standard conditions. The less carbon the better, since it collects in the ring grooves, causing sticking, and tends to shorten in their fuel than the small-bore high-speed types.

Ash content is determined by burning a fuel sample. Ash can abrade the rings and upper cylinder walls, and can attach itself to the nozzle tips, distorting the spray pattern.

The distillation temperature is a measure of how volatile the fuel is. Small engines run best on less volatile fuels.

## Combustion

For combustion we must have a fuel (which, in this case, is a hydrocarbon), oxygen, and heat. We obtain oxygen from the air. This odorless, colorless, and highly reactive gas composes about 21% of the atmosphere by volume. Heat is generated by compressing oxygen and nitrogen, as well as argon and the other trace constituents of air between the top of the piston and the roof of the combustion chamber.

Combustion does not take place immediately upon injection of the fuel. The individual fuel droplets vaporize on their surfaces. This vapor burns, and the process repeats itself until the droplet is consumed or, as is often the case, until the process stops because of oxygen starvation.

As a rule of thumb it takes about 0.001 sec for ignition to commence after the fuel is injected into the chamber. This may seem like a very short time, but for a high-speed engine, it is painfully slow. Relatively large quantities of vapor go off with a "bang," trapping the piston between skyrocketing pressures above and inertia below. This phenomenon is diesel *detonation*. It occurs *before ignition processes have rightly begun*, in constrast to SI detonation, which occurs after the spark. The effects—battered bearings and holed pistons—are the same.

## Ignition Lag

The graph in Fig. 3-2, adapted from the pioneering work of Sir Harry Ricardo, describes ignition delay, or lag. The vertical axis represents pressure, and the horizontal axis time for crankshaft degrees. Fuel is injected at A but does not ignite until B or the point at which total cylinder pressure exceeds compression pressure.

Ignition lag and consequent detonation can be combated in several ways. The simplest approach is to begin injection

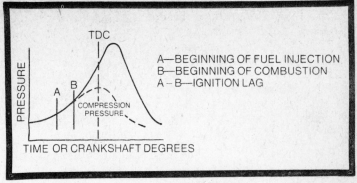

Fig. 3-2. Pressure rise diagram.

early, before cylinder temperature has reached the 500–750°F required for combustion. Mechanics have been known to use this expedient in the field when an engine crosses the threshold of detonation because of low-grade fuel dribbling nozzles, or some other short-term difficulty. A better solution is to raise the compression ratio. As the ratio goes up, the self-ignition temperature tends to drop, because oxygen is packed more densely around the fuel particles. Another design modification is to increase temperature of the charging air. Heat excites the gas molecules in the chamber and betters the chances of collision between the oxygen and fuel molecules. A similar effect can be achieved by increasing the coolant temperature, although there are obvious limits to this.

Turbulence is another way of reducing ignition lag. Either the charging air or the fuel spray can be made turbulent. In the case of the former, the inlet valve seats or ports can be angled to give a definite swirl to the air, or the combustion chamber can be designed to set up high-velocity currents. Turbulence can be induced in the fuel spray by careful design and positioning of the nozzles.

Inferior fuel—which, for a diesel engine, usually means that it has a high proportion of naphthenes and aromatics (precisely those families of hydorcarbons which are desirable for SI fuels)—can be *doped* to raise its cetane number. Ethyl nitrate, amyl thionitrate, and other compounds are used for this purpose.

As a practical matter it may be difficult to distinguish detonation from bearing noise, particularly with 2-cycles. Detonation generally occurs upon starting and under light loads—or, saying the same thing another way, when combustion temperatures are lower than normal. Knocks generally develop progressively over the life of the engine. As journal and piston skirt wear continue, knocks become louder and may be accompanied by low lube-oil pressure and increased lube-oil consumption.

Assuming the fuel has an adequate cetane number, sudden detonation can usually be traced to faulty injection. You can feel the pressure surge in the individual injector lines from a remote pump and you can hear the sharp click as the injector seats. If the feel of the line or the sound varies between cylinders, replace the injector with a known-good one. If this does not clear up the problem, assume that the pump is faulty.

# Fuel Systems

**4**

The fuel injection system has always been one of the difficulties in the CI engine. Fuel must be introduced against cylinder compression at some point before TDC (top dead center). Early engines employed air injection. A small amount of compressed air was mixed with the fuel and injected into the cylinder with it. Besides providing the energy for injection, compressed air helped to atomize the fuel. The disadvantages were several: The compressor—or, as was more often the case, the compressor stages—were parasitic loads, absorbing power without directly producing any. And the weight of the compressors, high-pressure plumbing, and starting tanks meant that the diesel was confined to stationary or shipboard applications.

In 1910 James McKechnie developed the first successful mechanical, or *solid* injector. Development continued in the next decade, with the Robert Bosch pump and injector as one of the most significant breakthroughs. Today all diesel engines employ some means of solid injection, although you may encounter a few ancient examples with the earlier system.

## FUEL INJECTORS

The fuel injector—variously called the *spray nozzle*, *spray valve*, or *fuel delivery valve*—is the final component of the fuel system and, in many ways, the most critical. The injector must deliver a timed and metered spray of fuel to the chamber, and must double as a check valve to prevent compression and combustion pressures from entering the fuel supply lines. In addition the injector is fitted with a return pipe to insure continuous fuel flow in the circuit.

The injector consists of two main parts—the injector body, and the nozzle and valve. (Fig. 4-1). The former mounts the injector on the cylinder head, carries the fuel inlet and spillage fittings and, in some designs, may incorporate a high-pressure pump. The nozzle and valve assembly are lifted from their seats when sufficient fuel pressure is developed.

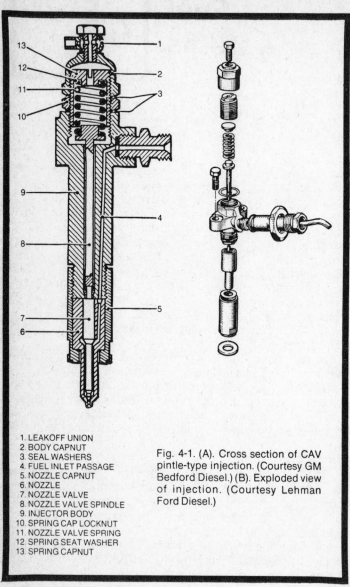

1. LEAKOFF UNION
2. BODY CAPNUT
3. SEAL WASHERS
4. FUEL INLET PASSAGE
5. NOZZLE CAPNUT
6. NOZZLE
7. NOZZLE VALVE
8. NOZZLE VALVE SPINDLE
9. INJECTOR BODY
10. SPRING CAP LOCKNUT
11. NOZZLE VALVE SPRING
12. SPRING SEAT WASHER
13. SPRING CAPNUT

Fig. 4-1. (A). Cross section of CAV pintle-type injection. (Courtesy GM Bedford Diesel.) (B). Exploded view of injection. (Courtesy Lehman Ford Diesel.)

The nozzle should:

1. Discharge cleanly, without "afterdribble."
2. Divide the liquid fuel into a fine spray for easy vaporization and combustion.
3. Direct the pattern deeply into the cylinder, but not against the piston or cylinder walls.

The spray pattern must have a definite start and stop to prevent fuel accumulations in the cylinder which would lead to detonation. The degree of atomization is a function of the viscosity of the fuel, fuel pressure, and of the ratio of orifice length to orifice diameter. For a fine pattern the orifice should be as direct as possible. Penetration depends upon fuel delivery pressure and upon the mass of the individual droplets. From the standpoint of penetration the droplets should be large. Since atomization and penetration are at odds, engineers try to work out a compromise which will adequately fill the cylinder but not induce too much ignition lag.

The simplest form of nozzle is the open-hole type. It consists of one or two check valves and an open orifice. The check valves open under fuel pressure and seat against cylinder pressure. These nozzles, with their single orifice, do not do much for turbulence and are only used in engines with some type of chamber-induced mixing.

The size of the hole determines the fineness of the spray. The smaller the diameter, the greater the atomization. But if the hole is too small it will not be able to deliver sufficient fuel. Multihole nozzles overcome this difficulty because the holes can be drilled small enough to atomize the fuel, and the number can be large enough to provide sufficient delivery. Depending upon the application, the nozzle may have as many as 16 orifices, with diameters as small as 0.006 in. The spray pattern is usually symmetrical, as shown in the inset to Fig. 4-2, although some chamber shapes and nozzle locations demand a canted pattern.

Some feature a pintle or pin which retracts into the orifice during discharge. The pintle is an extension of the nozzle valve. When retracted it releases a hollow, cone-shaped spray at an angle of up to 60° from vertical (Fig. 4-3). Pintle-type nozzles require, on the average, less maintenance than the multihole types, since the shuttle action of the pintle discourages carbon buildup and since the orifice diameter is larger.

A variant of this design is the throttling pintle, employed by Mercedes-Benz and others. The pintle has a tapered section. In the closed position the orifice is completely blocked,

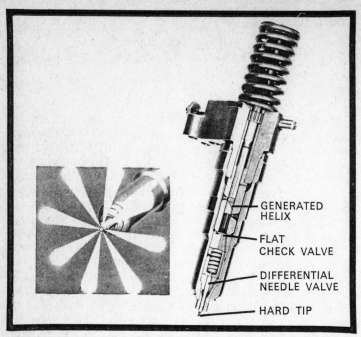

GENERATED HELIX

FLAT CHECK VALVE

DIFFERENTIAL NEEDLE VALVE

HARD TIP

Fig. 4-2. Multihole nozzle showing spray pattern. (Courtesy Murphy Diesel.)

and no fuel flows. Under part pressure, during the beginning of the pump stroke or when the distributor valve has just initiated delivery, the pintle retracts until its small section is in the nozzle orifice. It is unseated at this point, but fuel delivery is restricted by the diameter of the pintle. At full delivery pressure the pintle retracts clear of the orifice, allowing maximum fuel flow. Other throttling pintles work in the reverse manner—that is the pintle is fully retracted under zero or less than opening pressure, and is pushed outward as fuel pressure increases.

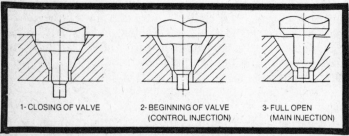

1- CLOSING OF VALVE

2- BEGINNING OF VALVE (CONTROL INJECTION)

3- FULL OPEN (MAIN INJECTION)

Fig. 4-3. Pintle nozzle action. (Courtesy Marine Engines Div., Chrysler Corp.)

## Fuel Delivery

The common- or third-rail system employs a continuously operating pump to pressurize the fuel header (Fig. 4-4). The individual injectors cam open in response to crankshaft rotation. Most common-rail systems include an accumulator in parallel with the header. The accumulator averages pump impulses and keeps a steady pressure in the system. Common-rail distribution is still in use, although it has certain disadvantages. It is a modification of air injection distribution systems, substituting pressurized fuel oil for air. However easy this may have made the transition some half-century ago, today it confuses mechanics rather than reassures them. More serious shortcomings are leaks, which seem to spring eternally in a pressurized header, and poor high-speed performance. Torque is a function of the weight of fuel consumed. Since the injectors are open for a shorter duration at high rpm, less fuel passes through them, and the engine develops correspondingly less torque.

The contemporary approach is to use *jerk pumps*. As the name indicates, these pumps pressurize the fuel intermittently and suddenly.

## Unit Injectors

Jerk pumps may be combined with individual injectors (Fig. 4-5). *Unit injectors*, as they are called, are energized by a camshaft or rocker arm. The nozzle opens in response to pressure developed in the pump assembly. These injectors are

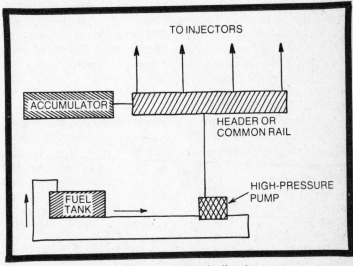

Fig. 4-4. Common- or third-rail system.

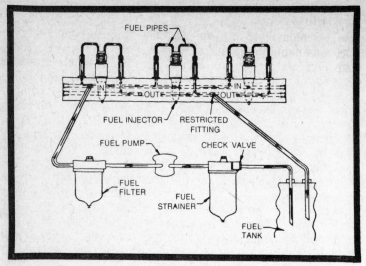

FUEL PIPES

IN — OUT — IN — OUT

FUEL INJECTOR — RESTRICTED FITTING

FUEL PUMP — CHECK VALVE

FUEL FILTER

FUEL STRAINER

FUEL TANK

Fig. 4-5. Unit injectors. (Courtesy Detroit Diesel Allison.)

found on many large engines and on GM's 2-cycle series. The most immediate advantage is the absence of high-pressure plumbing (fuel inlet pressure is 20 psi or so) and, hence, more reliability. Another advantage, at least for large engines, is that the designer can ignore pressure waves in the fuel column. When the injector opens, a low-pressure wave moves through the fuel on the discharge side of the pump. A high-pressure wave is generated when the pump column encounters the injector nozzle. Even open, there is enough of a restriction to send a high-pressure wave back to the pump at speeds approaching 5000 ft/sec. These waves can—depending upon the length and shape of the plumbing—reinforce each other to unseat the nozzle valve, thereby injecting fuel at the wrong time in the cylinder, or they can delay injection. The whole problem is bypassed by mounting the pump as a unit with the injector.

In addition to parts found on more conventional injectors, the unit injector has a mushroom-shaped cam follower, a return spring, a pump plunger, oil supply reservoirs, and one or more check valves. The pump plunger is moved up and down by the engine camshaft acting through push rods and rocker arms. The control rack is a bar with teeth milled on one edge that engage a gear on each injector. The rack is connected to the throttle and to the governor, and its position determines the amount of fuel delivered to the engine.

Metering is achieved by varying the effective stroke of the pump. The plunger reciprocates the same distance each time

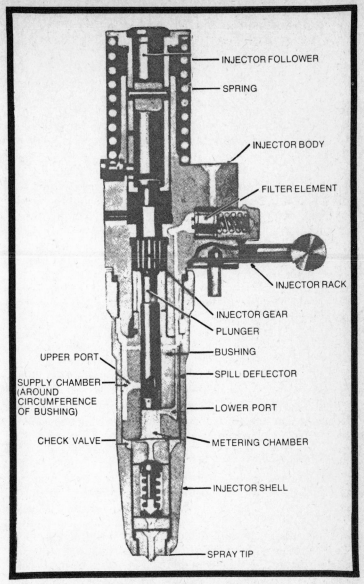

Fig. 4-6. Cutaway of GM-type unit injector.

it is tripped by the cam. But the effective stroke is a matter of the position of the fuel inlet and discharge ports relative to helical reliefs on the flank of the plunger.

Fuel enters the injector at the filter and passes to the supply chamber (Fig. 4-6). As the pump plunger moves down,

45

part of the fuel is routed to the supply chamber through the lower port, until the port is closed off by the edge of the plunger.

The plunger has an internal oil gallery that effectively shunts the pump as long as fuel trapped in the helix can escape. Its escape route is through the upper port, whose timing is a function of the contour of the helix. The lower port is blanked off by the plunger diameter once the downward stroke begins and plays no further part in the story.

With both ports closed, the fuel trapped below the plunger has no place to go except past the check valve and out the spray nozzle. The disc-shaped check valve, shown in profile in Fig. 4-6, is a fail-safe for the needle valve. Should the needle not seat because of dirt or carbon, the check valve will prevent air from entering the fuel supply.

The position of the control rack determines the timing of the upper port since rotating the plunger changes the contour of the helix. At no injection (rack full out) the upper port remains open until after the lower port is uncovered, as shown in Fig. 4-7. The pump idles uselessly, alternately delivering fuel through the lower port and then through the upper port. No pressure is developed below the plunger, and no fuel is delivered to the spray head. At full load the contour of the helix is advanced (rack full in), and the upper port is closed almost as soon as the lower port, giving the fuel no alternative but injection.

Part of the reason for the relative complexity of this particular unit injector is the need for lubrication and cooling. Oil is in constant circulation around and through critical parts.

## Remote Pumps

The injector pumps can be mounted remotely from the injectors and connected to them by high-pressure seamless tubing. Three configurations are used. In some of the larger engines a separate pump is used for each injector. The pump is a distinct entity, usually mounted on the head or the upper block. Another approach is to employ a single pump whose output is apportioned to the individual injectors by means of a distributor. Various fuel distributor schemes have been tried, but the simplest remains the most popular. It consists of a rotating plate with ports that are indexed with a second set on the cover. Figure 4-8 is a schematic of the CAV system, which combines high-pressure pump under the same cover as the fuel distributor, as well as the transfer pump, regulator, and governor. The major advantage of this arrangement is that there is no need to balance delivery to the various cylinders.

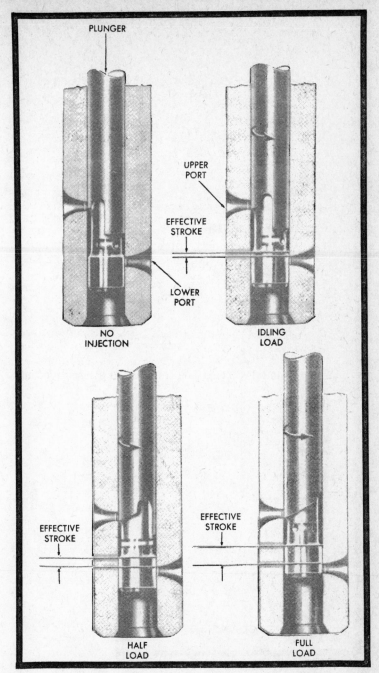

Fig. 4-7. Plunger positions from zero delivery to full load.

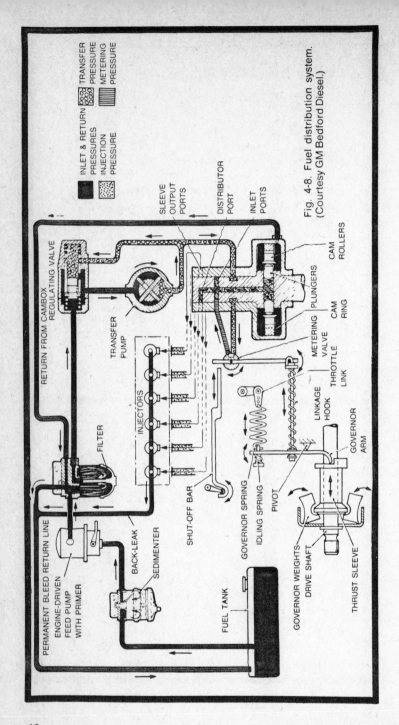

INLET & RETURN PRESSURES

INJECTION PRESSURE

TRANSFER PRESSURE

METERING PRESSURE

RETURN FROM CAMBOX REGULATING VALVE

TRANSFER PUMP

FILTER

INJECTORS

BACK-LEAK

SEDIMENTER

PERMANENT BLEED RETURN LINE
ENGINE-DRIVEN FEED PUMP WITH PRIMER

FUEL TANK

SLEEVE OUTPUT PORTS

DISTRIBUTOR PORT

INLET PORTS

CAM ROLLERS

PLUNGERS

CAM RING

METERING VALVE

THROTTLE LINK

LINKAGE HOOK

GOVERNOR ARM

GOVERNOR SPRING

IDLING SPRING

PIVOT

SHUT-OFF BAR

GOVERNOR WEIGHTS

DRIVE SHAFT

THRUST SLEEVE

Fig. 4-8. Fuel distribution system. (Courtesy GM Bedford Diesel.)

48

The same pump provides fuel for all of them. The CAV design is lubricated by fuel oil, and no external oil lines are needed.

## CAV Distribution System

The pump consists of a pair of opposed plungers which are actuated by cam rollers. On the inlet stroke the plungers move outward under pressure from the fuel delivered through the metering port (see Fig. 4-9). As the rotor turns, the metering or inlet port is blanked off by the rotor body. At the same time the plungers make contact with the cam lobes and are forced inward. Fuel passes through the distributor port, and to one of the injectors through the outlet port. There are as many inlet and outlet ports as there are cylinders. The amount of fuel injected per stroke is determined by the regulating valve (upper right in Fig. 4-8), which is controlled by the throttle and the governor. At low speeds the regulator reduces fuel pressure and, therefore, the amount of fuel delivered by the vane-type pump. Since the opposed pistons in the high-pressure pump are driven apart by incoming fuel, their outward displacement is determined by the amount of fuel passed by the regulating valve.

The contour of the cam provides relief of injector pressure near the end of the injector cycle, and prevents "afterdribble" at the nozzles. Timing of the individual cylinders is determined by accurate placement of the cam lobes and the outlet ports.

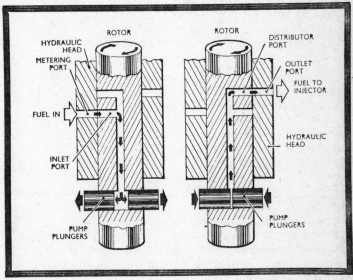

Fig. 4-9. Injector pump operation, distributor-type system. (Courtesy GM Bedford Diesel.)

The CAV system involves four pressures. Low pressure is generated by the lift pump, which delivers fuel to the transfer pump. Transfer pressure tends to increase with engine speed. The pressure curve is allowed to rise within strict limits imposed by the regulating valve. The fuel suffers a pressure drop as it enters the metering port on the distributor. The diameter of the orifice is tailored to different applications. Injector pressure is developed by the pistons. At idling speeds both transfer pressure and metering pressure are at their minimum value. Opening the throttle increases the amount of fuel delivered, and the engine accelerates to a speed corresponding to throttle position. Once this speed is reached the governor (which may be mechanical or pneumatic) holds transfer pressure constant.

## Bosch EP/VA Pump

The Bosch EP/VA series pump is, in some respects, more sophisticated than the CAV unit. It is intended to be used in passenger car service, which makes severe demands upon fuel metering and speed regulation. Figure 4-10 is a schematic of the main circuits. Other features such as the governor, hydraulic advance, and deferred idle are not shown.

Each upward stroke of the piston (shown at No. 1) involves three distinct events. The fuel transfer pump feeds the piston by means of two pipes, the back-pressure stage (2) and the regulator stage (3). The piston compresses the fuel and sends it to the injectors and high-pressure chambers (6) of the regulator shuttle (8). The compression chamber (2) is opened to the return circuit by the regulator shuttle, which moves to the right by virtue of regulator pressure and metering.

Metering is achieved by means of the shuttle. The main piston (1) moves upward, impelled by the pressure of the fuel passing through two parallel canals and emptying behind the shuttle. The shuttle moves to the right and controls the final injection by effectively varying the capacity of the compression chamber (2) and the diameter of the fuel-return port (11). On the downward movement of the piston the shuttle moves to the left under spring pressure (10) and closes the fuel-return port (11). Fuel is forced back through the two parallel lines connecting the piston and shuttle. The one-way valve (5) closes. The return of the shuttle spring pressure is a function of the opening of the accelerator valve (4). The amount of fuel injected is a function of the stroke of the shuttle, which, in turn, depends upon the position of the accelerator valve.

Compensation for load and fixed-throttle speed variations is automatic. As the speed increases the piston moves upward

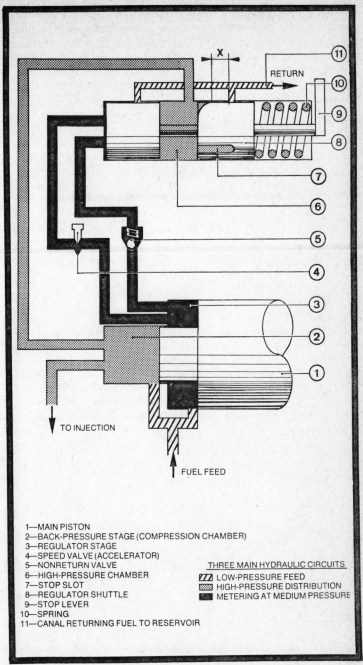

X

RETURN

TO INJECTION

FUEL FEED

1—MAIN PISTON
2—BACK-PRESSURE STAGE (COMPRESSION CHAMBER)
3—REGULATOR STAGE
4—SPEED VALVE (ACCELERATOR)
5—NONRETURN VALVE
6—HIGH-PRESSURE CHAMBER
7—STOP SLOT
8—REGULATOR SHUTTLE
9—STOP LEVER
10—SPRING
11—CANAL RETURNING FUEL TO RESERVOIR

THREE MAIN HYDRAULIC CIRCUITS

▨ LOW-PRESSURE FEED
▨ HIGH-PRESSURE DISTRIBUTION
▨ METERING AT MEDIUM PRESSURE

Fig. 4-10. Bosch pump schematic. (Courtesy Peugeot.)

51

more quickly. The shuttle moves back earlier, shortening the stroke and reducing the fuel volume per delivery. If the load increases the opposite happens, and more fuel is injected until new equilibrium is established.

Other distributor systems may be less elegant. Some act in conjunction with unit injectors to confine high pressure within the injector proper.

### Chrysler Marine Engine Pump

Another approach is to mount the pumps in a single unit a some distance from the injectors. Each cylinder has its own pump, but the pumps are ganged for compactness and drive

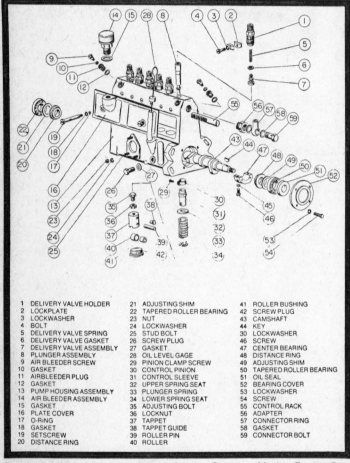

| | | | | | | |
|---|---|---|---|---|---|---|
| 1 | DELIVERY VALVE HOLDER | 21 | ADJUSTING SHIM | 41 | ROLLER BUSHING |
| 2 | LOCKPLATE | 22 | TAPERED ROLLER BEARING | 42 | SCREW PLUG |
| 3 | LOCKWASHER | 23 | NUT | 43 | CAMSHAFT |
| 4 | BOLT | 24 | LOCKWASHER | 44 | KEY |
| 5 | DELIVERY VALVE SPRING | 25 | STUD BOLT | 30 | LOCKWASHER |
| 6 | DELIVERY VALVE GASKET | 26 | SCREW PLUG | 46 | SCREW |
| 7 | DELIVERY VALVE ASSEMBLY | 27 | GASKET | 47 | CENTER BEARING |
| 8 | PLUNGER ASSEMBLY | 28 | OIL LEVEL GAGE | 48 | DISTANCE RING |
| 9 | AIR BLEEDER SCREW | 29 | PINION CLAMP SCREW | 49 | ADJUSTING SHIM |
| 10 | GASKET | 30 | CONTROL PINION | 50 | TAPERED ROLLER BEARING |
| 11 | AIRBLEEDER PLUG | 31 | CONTROL SLEEVE | 51 | OIL SEAL |
| 12 | GASKET | 32 | UPPER SPRING SEAT | 52 | BEARING COVER |
| 13 | PUMP HOUSING ASSEMBLY | 33 | PLUNGER SPRING | 53 | LOCKWASHER |
| 14 | AIR BLEEDER ASSEMBLY | 34 | LOWER SPRING SEAT | 54 | SCREW |
| 15 | GASKET | 35 | ADJUSTING BOLT | 55 | CONTROL RACK |
| 16 | PLATE COVER | 36 | LOCKNUT | 56 | ADAPTER |
| 17 | O-RING | 37 | TAPPET | 57 | CONNECTOR RING |
| 18 | GASKET | 38 | TAPPET GUIDE | 58 | GASKET |
| 19 | SETSCREW | 39 | ROLLER PIN | 59 | CONNECTOR BOLT |
| 20 | DISTANCE RING | 40 | ROLLER | | |

Fig. 4-11. Injector pump in exploded view. (Courtesy Marine Engine Div., Chrysler Corp.)

simplicity. Figure 4-10 illustrates a typical in-line injector pump of the type most used on small engines. The injectors, are operated hydraulically on cue from delivery pressure. Unlike unit injectors, there is no mechanical connection between them and the engine. The straightforward routing of the high-pressure tubing shown in the drawing is characteristic. It is achieved by indexing the pump cams with the firing order.

These pumps are, almost without exception, constant-stroke types. The individual plungers or pistons move the same distance with each revolution of the camshaft. Fuel delivery is adjusted by moving the control rack (No. 55 in Fig. 4-11) fore and aft. The rack rotates the plungers and varies the effective stroke. Each plunger has a helical slot running down its outer diameter. When the plunger is at the bottom of its stroke, fuel flows through both ports to fill the interior of the barrel, as shown in Fig. 4-12A. As the plunger is cammed upwards, some of this fuel is forced out of the ports until the plunger reaches the position at B. Further upward movement of the plunger increases the pressure on the fuel and causes the delivery valve to open, allowing fuel to enter the line connected to the injector. The circuit above the plunger is kept full of fuel by previous operations of the pump. The additional amount raises pressure and causes the injector valve to lift off its seat. Fuel is discharged into the cylinder.

Fuel discharge continues until the edge of the helix uncovers the spill port, as shown at C. Fuel in the barrel flows down the vertical recess, into the helix, and out the spill port. Once fuel can escape the barrel around the plunger, the effective stroke is over. Fuel delivery starts as soon as the plunger covers the spill ports, but is terminated by the helix—spill—port timing. When the port will be uncovered is a function of the rotation of the plunger, which is controlled by the position of the rack.

At C the plunger is shown in the full-load position, where it delivers a maximum fuel charge. Sketch D shows half-load,

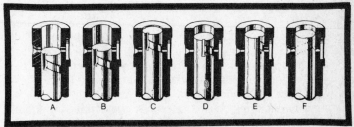

Fig. 4-12. Different strokes. (Courtesy GM Bedford Diesel.)

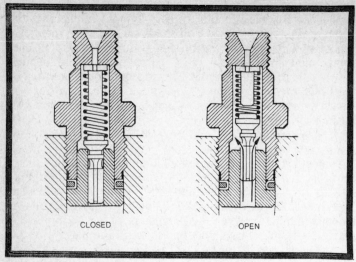

CLOSED            OPEN

Fig. 4-13. Fuel delivery valve. (Courtesy GM Bedford Diesel.)

and E shows idle. Drawing F represents shutdown; no fuel is delivered.

A variant is the Simms injector, which employs a helical relief in conjuction with a drilled hole along the plunger's axis. The configuration is reminiscent of the unit injector discussed earlier.

In some designs the injectors are not relied on to prevent afterdribble. A delivery valve (Fig. 4-13) is fitted at the top of each barrel. The valve unseats under pump pressure and closes when the spillway is uncovered. In closing, the valve draws a small quantity of fuel out of the pipe connected to the injector. This reduces the residual pressure on the pipe and hastens injector closing.

## FUEL SUPPLY PUMPS

High-pressure pumps have poor suction capability and must be primed by fuel supply or lift pumps. These positive-displacement devices operate more or less independently of fuel viscosity, pressure, or temperature. the geared type consists of a pair of meshed gears rotating inside of a housing. Pressure is derived from the action of the meshed gear teeth, which prevents oil from passing between the gears and, instead, forces it around the outside of each gear. These pumps generally include a pressure regulator in the casing.

Vane-type pumps consist of one or more pairs of sliding vanes in an eccentric housing. Pressure is developed as the vanes turn and wedge the fuel against the progressively

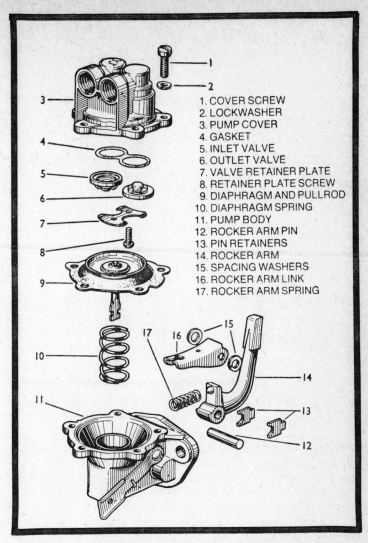

1. COVER SCREW
2. LOCKWASHER
3. PUMP COVER
4. GASKET
5. INLET VALVE
6. OUTLET VALVE
7. VALVE RETAINER PLATE
8. RETAINER PLATE SCREW
9. DIAPHRAGM AND PULLROD
10. DIAPHRAGM SPRING
11. PUMP BODY
12. ROCKER ARM PIN
13. PIN RETAINERS
14. ROCKER ARM
15. SPACING WASHERS
16. ROCKER ARM LINK
17. ROCKER ARM SPRING

Fig. 4-14. AC fuel supply pump. (Courtesy GM Bedford Diesel.)

narrower sides of the housing. Fuel is expelled through a discharge port on the periphery of the housing.

The most popular pump type is a diaphragm pump not unlike those used in automobiles. The diaphragm pump is by far the most popular for small engines. An AC design in shown in exploded view in Fig. 4-14. It is driven from the engine cam or from a lobe on the injector pump cam. Note the provision for manual priming.

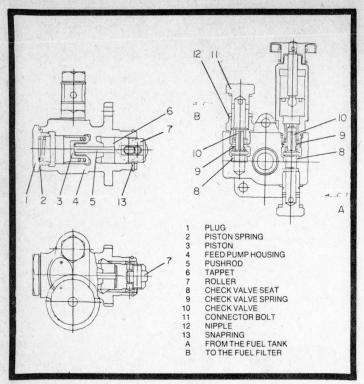

| 1 | PLUG |
| 2 | PISTON SPRING |
| 3 | PISTON |
| 4 | FEED PUMP HOUSING |
| 5 | PUSHROD |
| 6 | TAPPET |
| 7 | ROLLER |
| 8 | CHECK VALVE SEAT |
| 9 | CHECK VALVE SPRING |
| 10 | CHECK VALVE |
| 11 | CONNECTOR BOLT |
| 12 | NIPPLE |
| 13 | SNAPRING |
| A | FROM THE FUEL TANK |
| B | TO THE FUEL FILTER |

Fig. 4-15. Piston-type transfer pump. (Courtesy Marine Engine Div., Chrysler Corp.)

Figure 4-15 illustrates the piston pump used on Chrysler-Nissan SD22 and SD33 engines. It is integral with the high-pressure pump and driven from the same cam. The operation of the pump is slightly unorthodox in that fuel is present on both sides of the piston, thus eliminating the possibility of air lock.

Figure 4-16 diagrams its operation. On the downward movement of the piston the discharge-side check valve (No. 3) closes, and fuel goes up the outer chamber (8) to the high-pressure pump. The intake-side valve (No. 4) is open. At position II the piston is forced up by the rotation of the cam. Fuel enters the outer chamber (B) through the check valve (3) and the passage on the intake side. Discharge commences at position III.

**FUEL FILTERS**

Fuel oil doubles as a lubricant for the pumps and injectors and must be compatible with parts whose tolerances are gaged

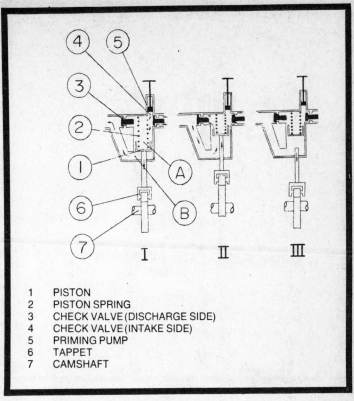

| 1 | PISTON |
| 2 | PISTON SPRING |
| 3 | CHECK VALVE (DISCHARGE SIDE) |
| 4 | CHECK VALVE (INTAKE SIDE) |
| 5 | PRIMING PUMP |
| 6 | TAPPET |
| 7 | CAMSHAFT |

Fig. 4-16. Pump operation. (Courtesy Marine Engine Div., Chrysler Corp.)

in hundred-thousandths of an inch. Refiners take pains to keep diesel oil free of impurities, but contaminants can enter the fuel at any point from distillation to final tankage. Filters are vital to efficient and economical operation (Fig. 4-17).

Most engines incorporate at least three stages of filtration. The filters are the full-flow type, in series with the fuel line. A bypass valve is usually present to keep the fuel flowing if the filter clogs. In many applications it is better to risk scoring the pump than to suffer unexpected shutdown. The filter stages are progressive: The primary stage is usually rather broad-gaged and traps particles larger than 0.005 in. or so. The secondary stage, which is located between the supply and the injector pumps, has a pleated paper element which filters down to 25 microns (0.0009 in.) or finer. Where water is a problem, as in marine or occasional-use applications, a separator is fitted. It may be a glass or aluminum bowl upstream of the supply pump, or it may be integral with the

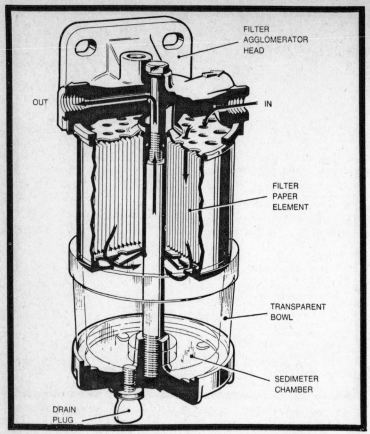

FILTER
AGGLOMERATOR
HEAD

OUT

IN

FILTER
PAPER
ELEMENT

TRANSPARENT
BOWL

SEDIMETER
CHAMBER

DRAIN
PLUG

Fig. 4-17. Filter cutaway showing fuel flow. (Courtesy GM Bedford Diesel.)

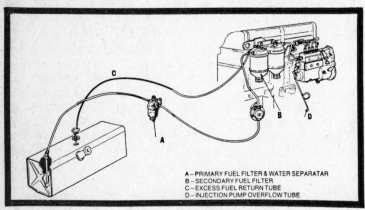

C

B

D

A

A – PRIMARY FUEL FILTER & WATER SEPARATAR
B – SECONDARY FUEL FILTER
C – EXCESS FUEL RETURN TUBE
D – INJECTION PUMP OVERFLOW TUBE

Fig. 4-18. Typical fuel system. (Courtesy Lehman Ford Diesel.)

primary filter. In addition, some engines feature filters at the injectors to give added protection to these sensitive components. Figure 4-18 illustrates a typical fuel supply system.

Filter casings usually have bleeder valves mounted high to purge the air from the lines, and many also incorporate return valves to divert excessive fuel back to the tank (Fig. 4-19).

### Fuel System Refinements

The basic fuel system—injectors, high-pressure pump, transfer pump, and staged filters—has been described. As fuel costs have climbed and diesels have moved into territory hitherto occupied by SI engines, certain refinements to the fuel system have been introduced. The most significant of these are intended to improve the engine's flexibility at either extreme of the rpm scale.

### Fuel Timer

Many current designs employ a fuel timer such as the one shown in Fig. 4-20. The timer is integral with the injector pump and operates to advance the pump cam at high speeds. Ignition delay is relatively constant. That is, it takes about as long for the fuel to ignite at 4000 rpm as it does at 1000, although the turbulence at high speed accelerates combustion to a minor extent. At high speed, lag becomes significant because the engine sees it in terms of crankshaft degrees. The timer compensates by introducing the fuel a few degrees early.

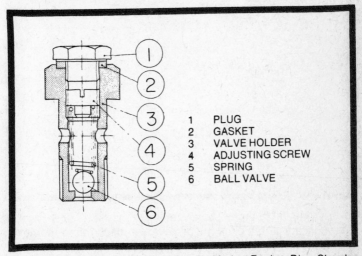

| | |
|---|---|
| 1 | PLUG |
| 2 | GASKET |
| 3 | VALVE HOLDER |
| 4 | ADJUSTING SCREW |
| 5 | SPRING |
| 6 | BALL VALVE |

Fig. 4-19. Fuel overflow valve. (Courtesy Marine Engine Div., Chrysler Corp.)

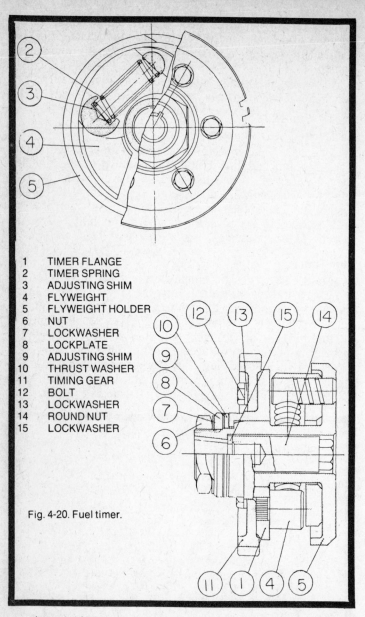

1 TIMER FLANGE
2 TIMER SPRING
3 ADJUSTING SHIM
4 FLYWEIGHT
5 FLYWEIGHT HOLDER
6 NUT
7 LOCKWASHER
8 LOCKPLATE
9 ADJUSTING SHIM
10 THRUST WASHER
11 TIMING GEAR
12 BOLT
13 LOCKWASHER
14 ROUND NUT
15 LOCKWASHER

Fig. 4-20. Fuel timer.

A typical timer is shown in two views in Fig. 4-20. It is reminiscent of the centrifugal advance mechanism employed on ignition distributors. As engine speed increases, the flyweights pivot outward against their restraining springs. This movement is transmitted to the pump camshaft and turns the split shaft against the direction of its rotation. The pump discharges earlier.

The timer should be checked for freedom of movement and maximum advance. The model shown here is allowed 5–7.5° advance, depending upon application. Travel limit is determined by shims.

## Peugeot Deferred Injection

*Deferred injection* is the term Peugeot uses for its patented delay system. Diesel engines tend to knock and rattle at idle because of injection delay. Actually the knock is present at higher speeds, but is blanketed by mechanical and exhaust noise. A cold engine is characterized by a more marked and randomized delay and is consequently quite rough.

Engineers at Peugeot have at least solved part of the problem, although the 504 still bucks and humps on a cold morning. The deferred-injection device is in a unit with the fuel distributor and acts in conjunction with an accumulator to increase the duration of fuel discharge at idle. By increasing the duration less fuel is injected into the cylinder during any time period. When the mixture finally does ignite, pressures are lower, and noise and vibration are reduced.

The drawings in Fig. 4-21 help to clarify the operation of the device. Fuel is pressurized in the chamber (Fig. 4-21, No. 7) and is pumped through the injector discharge port (5) and through a line to the accumulator (6). Only one delivery valve is shown, for simplicity. There are four of them—one for each cylinder, threaded into the hydraulic head. Part of the fuel passes through the open valve (4) of the accumulator, into the chamber (3), and lifts the piston (2) against its spring (1). During injection the accumulator continues to collect some fuel. The effect is to spread out delivery over time by absorbing the sudden pressure rise generated by the pump.

## FUEL SYSTEM MAINTENANCE AND REPAIR

The fuel system is most critical from the point of view of preventive maintenance. Failure to follow a maintenance and inspection schedule will result in time wasted in shutdowns, and can cause expensive repairs. Filter elements are cheaper than injector pumps.

### Bleeding and Priming

Whenever the system has been opened or whenever the tank has run dry, the system must be bled. Some fuel will escape in the process. Place a drip pan under the engine and be sure that the wiring harness is clear of the spillage. Fuel oil softens insulation.

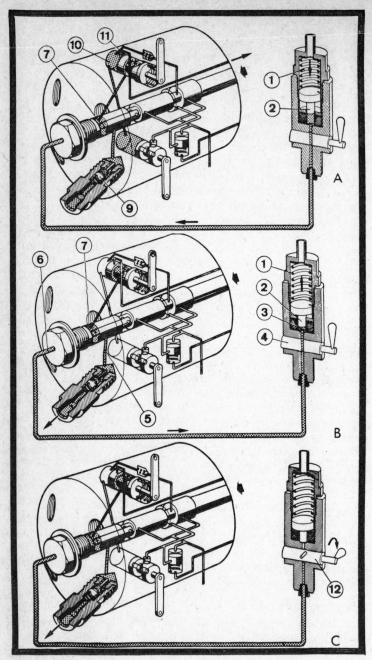

Fig. 4-21. Peugeot deferred-injection device. (**A**). Action during injection.
(B). Action after injection. (C). Action with accelerator depressed.

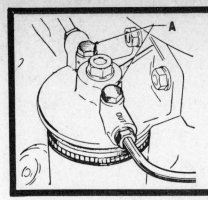

Fig. 4-22. Fuel filter bleed screws. (Courtesy Lehman Ford Diesel.)

Bleeder screws are located in the upper filter bodies (Fig. 4-22) and in one or more locations on the accessible side of the injector pump (Fig. 4-23). Carefully clean the screw tops and adjacent areas to prevent foreign matter from entering the fuel supply. A hand-operated priming pump is incorporated with the injector pump or integral with the transfer pump. Working from the tank end forward, crack the bleeder screws, and tighten as fuel flows in an unbroken stream. It may be necessary to break the high-pressure lines at the injector pumps. Continuing to keep pressure up (10 psi is adequate), crack one union at a time, and hold until the bubbles disappear. Run the engine for at least 10 min. to insure that the fuel system is completely purged.

### Filter Service

Sediment chambers should be drained daily or prior to startup. To replace the filter, drain the fuel out the petcock.

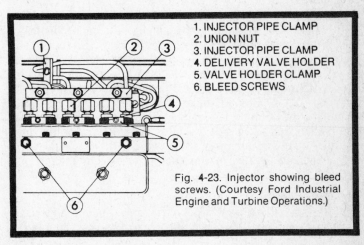

1. INJECTOR PIPE CLAMP
2. UNION NUT
3. INJECTOR PIPE CLAMP
4. DELIVERY VALVE HOLDER
5. VALVE HOLDER CLAMP
6. BLEED SCREWS

Fig. 4-23. Injector showing bleed screws. (Courtesy Ford Industrial Engine and Turbine Operations.)

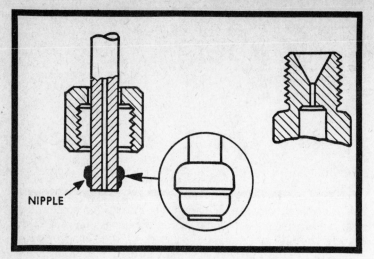

Fig. 4-24. Inspect the nipple (ferrule) and fitting for concentric wear patterns. (Courtesy GM Bedford Diesel.)

Remove the mounting bolt. Clean the shell in fuel oil and dry with a lintless rag or, preferably, compressed air. Density-type elements should be immersed in fuel oil prior to installation, to expel trapped air and make starting easier. Fill the shell to about two-thirds capacity with fuel oil. Install new gaskets at the recess and, if present, at the union. Tighten the cover nut just enough to prevent air leaks. Do not jam the nut since this might distort the shell or cover. Bleed the system and start the engine.

### Fuel Lines

Inspect the high-pressure fuel pipes and fittings carefully (Fig. 4-24). Overtightening the unions will cause distortion which may be seen on the pipe ferrule (flange). Deep scratches or abrasions can develop into cracks. Replace any pipes and fittings with these defects. When installing, tighten finger tight and check that the pipe is centered as it emerges from the union. Firm up and check the lay of the lines. They should not be in contact with the block or any other part. When retaining clips are used, the clips should fit over the lines without straining them. In other words, there should be no pretension on the lines other than that offered by the terminal fittings.

### Transfer Pump

Periodically (every 200 hours or as per maintenance schedule) remove the cover and clean the fuel chamber of diaphragm pumps (Fig. 4-25). Examine the diaphragm for tears, cracking, and loss of elasticity. Piston- and gear-type pumps do not require routine inspection and should not be disturbed short of failure or overhaul. In most instances failure will come about because of a clogged valve. The valve may be lapped to fit, or, if that is impractical, replaced. GM rotary pumps may fail at the drive shaft or coupling, though this is rare. You can determine if the pump is turning by inserting a fine wire through one of the drain ports. The wire will vibrate when struck by the gears.

Pump delivery rate and pressure specifications are not always available. You can get a rough idea of pump efficiency by disconnecting the line between it and the high-pressure pump. Collect the spillover in a glass jar and look for bubbles which indicate that the pump is losing its prime by sucking air. If this is the case, check all fittings and lines on the suction side of the pump. Pay particular attention to strainers or filters between the pump and tank.

### Injectors

Injectors, like spark plugs, are to be assumed guilty until proven innocent. Injector failure has many symptoms, some of which are more pronounced for the different engine brands and models. You may encounter one or more of the following symptoms:

- Black or gray smoke: If the smoke issues in puffs, one injector/pump is involved.
- White smoke: evidence of one or more cylinders cutting out.
- Detonation.

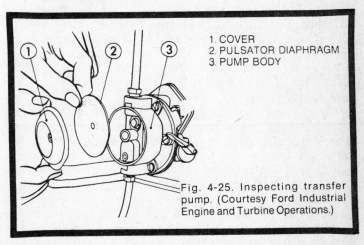

1. COVER
2. PULSATOR DIAPHRAGM
3. PUMP BODY

Fig. 4-25. Inspecting transfer pump. (Courtesy Ford Industrial Engine and Turbine Operations.)

- Uneven running, or stalling under load: This symptom could point to several faulty injectors, although for engines which have been properly maintained, one would first investigate the fuel system below the injectors.

Identifying the particular injector or injectors at fault is not difficult. Run the engine at idle. If it is fitted with unit injectors, depress one follower at a time with a screwdriver blade. If the injector in question was functional, disabling it will cause an rpm drop and increased engine roughness. Pump-fed injectors may be isolated by cracking their fuel delivery lines.

Some injectors require a special tool or a modification of an existing tool for removal. Disconnect the fuel lines and locating clips (if so fitted). Pull the lines clear, being careful not to bend them. Cap the lines with shipping caps. Removal of unit injectors involves removal of the associated rocker arms and the fuel rack.

The injector should be checked with the appropriate tester or, lacking that, checked against a known-good one fitted to the engine. Clean the injector sleeve (the part which mates the injector body to the head) to remove all traces of carbon. Failure to do this may cause injector overheating. Reamers are available, but you can do almost as well with a rotary wire brush chucked in a drill motor. Coat the sleeve with grease to help contain the carbon.

In working on injectors, *cleanliness* is the watchword. The work environment must be clean almost to the point of being antiseptic.

Use new gaskets. On injectors with two bolt mounts, it is a good idea to thread the fuel lines before the injector is tightened. This precaution will stress the lines evenly. Tighten the injector to specifications and crank the engine with the control rack pulled out. If there is any leakage around the joint, loosen the mounting fasteners and retorque. Run the engine and observe the fuel line fittings. Should they leak, loosen and retighten.

The injector should be tested with pressure gage and hand pump. The test apparatus is available from engine and accessory manufacturers. When using it observe reasonable safety precautions: the fuel spray will easily penetrate the skin and cause serious and difficult-to-treat infections. Do not pump the device beyond its rated capacity.

Mount the injector in the fixture provided for it and place a pan under the tip to contain the spray. Fill the tester reservoir with diesel oil (a few engine builders specify kerosene) and cycle the injector a few times to purge the lines. Slowly come up to pressure and note the gage reading when the injector opens. It should be the same for all cylinders and within the manufacturer's specs. Low opening pressure is often caused by a weak spring, which can be shimmed (Bosch) or tightened by means of a screw (CAV). Other wear on the needle valve may also cause early opening. Excessive pressure requirements can usually be traced to carbon accumulations at the tip.

The injector should hold 20% or so of opening pressure without dribbling fuel from the tip. Leakage means that the internals must be cleaned and lapped.

The spray pattern is critical. Pintle-type nozzles typically give the pattern shown in Fig. 4-26.

### Fuel Injection Pumps

Pump failure can be distinguished from injector failure by substitution of known-good injectors. Pump efficiency—a function of rpm and delivery rate—can only be determined with the aid of a power-driven calibrating machine. These machines are expensive and beyond the pale of the average shop.

Prior to removal, scrub the pump body and mounting-flange area with solvent and blow dry. Bar the engine over to identify the timing marks for refitting. Remove the fuel lines and oil lines (if fitted). Remove the pump mounting bolts and lift clear. Cap the fuel lines and clean the

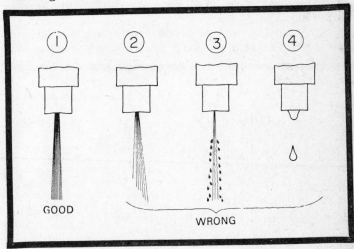

Fig. 4-26. Spray patterns for pintle-type nozzle. (Courtesy Marine Engine Div., Chrysler Corp.)

pump casting a second time, now that all surfaces are accessible. Remove the subassemblies, which may include the drive coupling, transfer pump, fuel timer, and associated filters. At this point the pump may be sent out for overhaul and calibration. Pump work is the most demanding of all diesel maintenance and should be done by experienced, knowledgeable mechanics who have a feel for precision parts.

The disassembly drill for the pump proper varies with makes and models. Normally, the first step is to remove the delivery valves, marking them or arranging them on the bench to insure proper assembly. Rotate the cam so that each tappet is in its *up* position. Shim the tappets and withdraw the cam. The tappets are removed from below. The tappet bores may be covered by the lower plate or by threaded plugs. Mark or arrange the tappets sequentially. Withdraw the plungers from the top with the aid of a length of hooked wire or expansion forceps. *Do not touch the lapped surfaces with your fingers.*

Inspect the plungers and bore for wear with a powerful magnifying glass. Scratches deep enough to be felt mean that the plunger must be replaced. This kind of wear comes about because of poor filter maintenance. Dulled rubbing surfaces are caused by water or acid contamination. Polish lightly to restore the sheen. Uneven wear patterns indicate that the bore has been warped from overtightening the fuel delivery valve. Wear in the helix profile destroys pump calibration.

A plunger should fall of its own weight when the barrel is held 45° from vertical. Gummed or varnish-covered plungers can make governor action erratic and accelerate wear on the helixes and rack.

New plungers must be handled with the greatest care. Soak in kerosene or other solvent to dissolve the preservative,

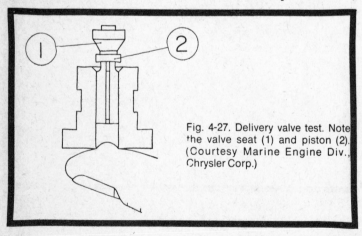

Fig. 4-27. Delivery valve test. Note the valve seat (1) and piston (2). (Courtesy Marine Engine Div., Chrysler Corp.)

and dip in fuel or test oil. Assemble without leaving fingerprints on the lapped surfaces. Be sure that the plunger mates with the control mechanism.

Clean the delivery valves and make a functional test by depressing the plunger and releasing it as shown in Fig. 4-27. It should hold air pressure. Most valves cannot be resurfaced and must be replaced as an assembly.

With a micrometer, check the cam against the original specifications. Inspect the control rack for wear. Figure 4-28 illustrates, in exaggerated fashion, the knife-edged wear pattern. Check the rack bushings for wear and ovality by fitting a new control rack. Comparison of the sliding fits will show the state of the bushings. New bushings must be reamed to fit.

Tappet height fixes the stroke of the pistons and so determines, to great part, the pump output. Initially all tappets should be set to the same height, either by means of shims or integral adjustment screws. Before the pump is installed on the engine it should be calibrated and the individual elements phased. This job requires a machine sensitive to cam angle and to output.

Assemble the pump with new gaskets, oil seals, O-rings, torquing the fuel delivery valve holders and other critical

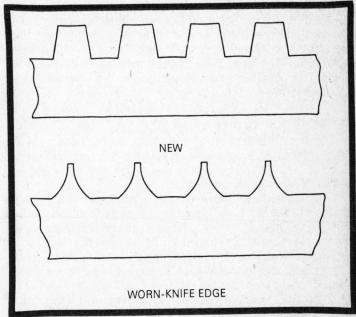

NEW

WORN-KNIFE EDGE

Fig. 4-28. Control rack weak. (Courtesy GM Bedford Diesel.)

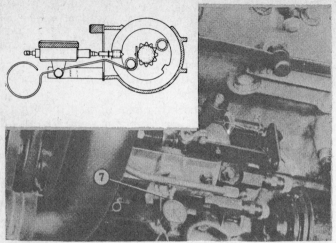

Fig. 4-29. Mounting pump with the aid of a dial indicator. (Courtesy Peugeot.)

parts to spec: Some manufacturers use adhesive in conjunction with the gaskets to make a secure joint. Scrape the old joints clean and use the brand of adhesive originally specified.

Assembling the pump to the engine can be complicated by the need to align timing marks or other reference points. The Peugeot requires special tools. Figure 4-29 shows a dial indicator and mounting fixture and (see inset) backlash takeup tool. The latter is the hooked wire bearing against the pinion. The idea is to mount the pump in such a position that all backlash is taken up.

## Timing

Timing drills vary all over the map, but the purpose of the exercise is to sychronize No. 1 plunger delivery with No. 1 cylinder. Once this is done the other plungers and cylinders will automatically fall into time.

All engines have two reference marks. One mark is on the flywheel or the harmonic balancer and refers to piston position relative to top dead center. The other mark is implicit in the pump and is usually stamped in plain view. It refers to the position of No. 1 plunger at the moment of fuel delivery. These two marks have to be aligned so that they both pass their pointers at the same instant.

Mercedes-Benz has been somewhat coy in this regard. A pump mark is not used. Instead one substitutes a length of clear plastic line for No. 1 fuel pipe and bars the engine until the fuel level rises in the plastic tube. As soon as the fuel moves we know that No. 1 plunger has come into play. Bedford engines are more typical in that they have flywheel marks (visible through a window) and marks on the pump drive coupling. One turns the engine until the flywheel marks align, and then slacks the coupling to match those marks. International Harvester engines are timed before the pump is installed. The drive gears have match marks. Do not be confused with the marks on the idler gear—they are not significant for this particular operation.

The surest and fastest way to time an engine is with a diesel timing light. The *Sun* brand is representative of the best. The advantage of a light is that it is triggered by fuel line pressure and so gives a direct reading. The fuel injection pressure wave travels rapidly through the oil column in the line. And depending upon engine speed and pump-to-injector distance, injection can be delayed by as much as 9° of crankshaft rotation. Engine makers are well aware of this phenomenon and attempt to compensate for it by advancing the pump timing marks or by calling for more advance than the engine needs or will get. The light tells the truth and will check the operation of the fuel timer.

Figure 4-30 shows the hookup. A transducer is installed at No. 1 plunger output. Certain engines, such as the 1100 series Caterpillar—Ford engine require that the transducer be installed at the injector. Power is by way of the engine's own 12 or 24V batteries or from a 115V household-power line. The engine is idled and the strobe is pointed at the crankshaft timing marks. The dial on the back of the light will show the number of degrees of pump advance before TDC. To check the automatic advance, the engine is speeded up. In addition, the dial carries an rpm calibration, so that one man can set the governor. No longer does someone have to stand around holding a tachometer against a crankshaft pulley or other revolving part.

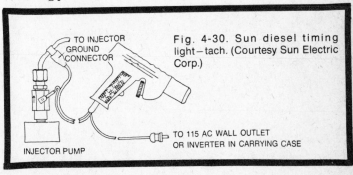

TO INJECTOR
GROUND
CONNECTOR

INJECTOR PUMP

TO 115 AC WALL OUTLET
OR INVERTER IN CARRYING CASE

Fig. 4-30. Sun diesel timing light—tach. (Courtesy Sun Electric Corp.)

# Governors 5

The majority of diesels employ centrifugal flyball governors which are integral with the injector pump. Operation differs little from the flyball governor invented by James Watt for his steam engines. But instead of opening and closing a steam valve, the diesel governor rotates the injection pump plungers to deliver more or less fuel per stroke. Pneumatic governors respond to air velocity entering the manifold, which, in turn, is a function of piston speed.

No governor can act instantaneously—the engine senses and responds to a load before the governor can react. Coarse regulation is adequate for most installations and may result in speed change peaks of 10% or so. Fine regulation cuts that figure in half, for maximum of 2½% over or under the desired rpm.

Service work on governors consists of adjustment (particularly of the high- and low-speed limits), cleaning, and parts replacement. Bearings and pivots wear, coarsening the regulation. And occasionally a diaphragm will fail because of leaks or age hardening.

## CENTRIFUGAL GOVERNORS

Before any work more demanding than adjustment is undertaken, you should try to understand how the mechanism operates. We will consider two rather complex governors in this section. They are similar to others in all but small detail. The trick is to concentrate on what typewriter mechanics call the *chain of motion*, or how movement of one part is transferred to others.

### CAV Governor

The CAV governor is quite conventional, but it does have a manual override, which complicates matters a bit. The lever

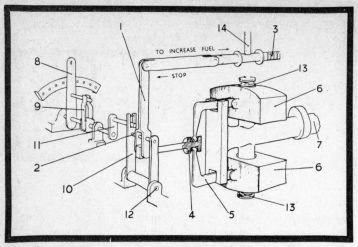

Fig. 5-1. CAV midspeed setting. (Courtesy GM Bedford Diesel.)

(*1* in Fig. 5-1) pivots on the axle (*2*). The whole assembly pivots at point *12* on command from the throttle (*8*).

In Fig. 5-1 the engine is running at about half-speed. A change in speed will be reflected by the flyweights (*6*). Should the engine accelerate, the weights will press outward against their springs (*13*) and move the lever and rack to the right, reducing fuel delivery. Under load the engine slows, and the weights, impelled by their springs, move inward, causing the

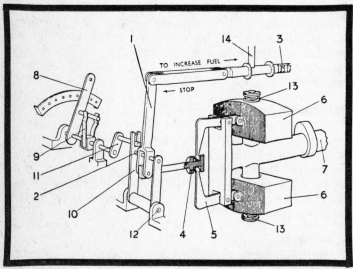

Fig. 5-2. CAV three-quarter-speed setting. (Courtesy GM Bedford Diesel.)

throttle to open. Now let us increase speed by moving the control level (8) a few clicks to the right, as shown in Fig. 5-2.

The yoke (10) pivots on its axle (12), thus moving the lever (1) closer to the flyweights. To keep the geometry constant, the lever is kicked to the right by the spool shaft (4). Consequently, more fuel leaves the pump, and the engine accelerates. When centrifugal force is balanced by the counterweight springs, speed will stabilize, but at a higher rpm than would be the case in the previous drawing.

It is possible that the rack may be opened until it contacts its stop. Additional movement of the control lever would be only possible by forcing the weights apart. Rabbit-ear springs (9) are provided to protect the mechanism.

To reduce the speed the control lever is moved to the left, and it brings the lever (1) and yoke (10) with it. Regardless of engine speed the weights describe the same track.

**CAV Repairs.** Governor malfunctions—hunting, sticking, refusal to hold adjustments—can usually be traced to binding pivots. Most of the time the pivots bind because of dried oil and dirt accumulations. After long service the pivots and bearings may wear enough to require replacement. The drawing in Fig. 5-3 shows most of the 200-odd parts which make up a CAV governor, and can serve as our reference in this discussion. The bearings to check are numbered as 223, 224, 225, and 226. Check the bushings on the bellcrank levers (223 and 224) and the element support sleeve (231). Bushings should be pressed into their bosses and reamed. Use a factory-supplied reamer for best accuracy.

The pivot pins, shown at bottom right as 234, fit into bores on the sides of the weights. Check these bores for wear. Weights are sold in matched pairs.

Further disassembly requires that the oil be drained and all external connections opened. Remove the filler and oil level plugs (144 and 161). You will observe two hex nuts (191) secured with safety wire. Cut the wire and remove the nuts. The timing lever and fulcrum pin (190) can now be withdrawn. Next, remove the pair of nuts (159) which hold the control lever (200) and flange assembly (193) in place. Remove the assembly from the end cover.

The end cover is held by eight screws. Once they're off you will be able to get to the vitals of the governor. At this point disassembly becomes complex and promises that assembly will be equally so. Lay the parts out on a paper-covered bench, in rows from left to right. When you put the mechanism back together, follow the pattern exactly as if you were reading a newspaper.

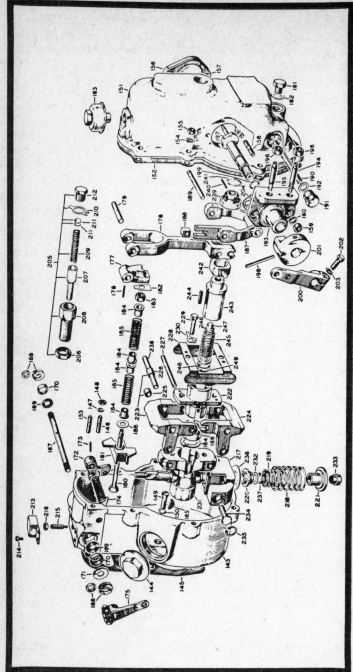

Fig. 5-3. CAV governor in exploded view. Refer to the text for parts idention. (Courtesy GM Bedford Diesel.)

The control rod linkage (177) is unscrewed from the control rod, and the assembly withdrawn by rotating it 90° and disconnecting the yoke-shaped lever. The governor weights carry two screws (229). Remove them and then observe that the plate (228) is secured by a pin. Rotate the shaft so that the pin can be punched free. The governor linkage now can be separated from the weights.

Next, remove the guide bush (222) from the end of the camshaft. The bush floats freely and should lift right out. Behind it you will find the governor weight assembly nut (166). You will need a CAV tool to remove this nut, which can be ordered from your distributor as part number 7044-112B. Another tool, No. 7044-8, is needed to pull the weights. Bottom the puller in the threads before applying force; otherwise, the threads might strip and you will be in real trouble.

Remove the key (150) from the camshaft keyway and the linkage at the control rod. The rear cover (151) is next, along with the auxiliary idle stop (205). Under the plug (212) lurks a spring (209) and a plunger (207); it is rarely necessary to disturb these parts.

The weights are usually treated as an assembly. If you need to replace them (as in the case of worn pivot holes), place the weights in a brass-jawed vise and compress the spring by exerting force on the upper spring plate (221). Remove the adjusting nut with a CAV No. 7044-65 on each weight. Remove the spring clips (235) and their pins (234). You can now lift off the two bellcranks (223 and 224). The next step is to remove the excess-fuel hood (213), the excess-fuel adjusting screw (215), spindle stop lever (175), and the spring clip (168). Replace this clip with a new one on reassembly. The pawl (172) is located by a pin (173). Drive the pin clear and withdraw the spindle from the housing. Discard all seals.

**CAV Assembly.** Begin assembly with the bellcranks (223 and 224), using the pivot pin (236) to locate them on the element support sleeve (231). Fit the weights (217) to the bellcranks with the pins (234) and fit the pin clips (235). Replace the excess-fuel-device seals (170) and assemble the pawl (172) and pawl spring (174). Hold them in position while you fit the spindle (167) into the housing. Pawl and spindle go together with a pin (173). With new clips on the spindle, mount the control lever (175). All parts should move freely at this and subsequent stages of assembly. If something binds, find out why before you proceed.

Fit the control linkage to the control rod and tighten the tab. Next, insert the key (165) on the camshaft and fit the weight assembly over the shaft. The spring washer and nut should be tightened to 500 in.-lb with CAV No. 7044-112B. The brass element guide (222) has its slotted end nearest the pump camshaft. Screw the control link assembly onto the control rod. Rotate the governor weight assembly and camshaft to give access to the bellcrank pin boss. Insert the pin (227) and secure with the retaining case (228). Tighten the screws and secure with the locking tabs provided. With a new gasket on the rear housing cover, assemble and tighten the eight nuts in a crisscross progression. The timing lever and fork pivot pin (190) go through the lower pair of holes in the rear cover. Use new sealing washers (192). The capnuts are threaded to the timing lever and fork pivot, and not to the housing cover.

Fit the lever and flange assembly (193) with a new gasket. Note the way in which the eccentric on the camshaft adjusting lever (197) fits into a hole on the block (188). Tighten the fasteners, being sure to include the lockwashers. Finally, replace the oil filler plug and drain plug. Use a new gasket on the drain. Mount on the engine and adjust as required.

## Simms Governor

Although the Simms governor is a bit simpler than the CAV, it operates on the same principle; a force acting to pivot the weights outward is balanced by springs holding the weights to the shaft. Centrifugal force acts to pivot the weights outward in proportion to shaft speed. This force is

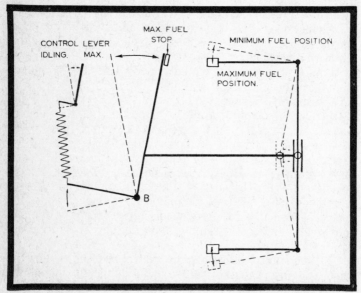

Fig. 5-4. Simms governor action. (Courtesy GM Bedford Diesel.)

counteracted by spring tension. An override is provided for throttle control at speeds above idle and below maximum.

Figure 5-4 is a schematic of its operation. At start, the fuel rack is all the way in. (The control lever position is shown as the solid line.) The governor weights will be pressed inward. When the engine starts and picks up speed, the weights are flung outwards and cam against the hub sleeve, moving it horizontally. The hub sleeve moves the control lever to the position indicated by the dotted lines in the drawing. The rack will be moved toward the "no fuel" position. To avoid abrupt movement, Simms engineers have fitted two springs between the crank lever and throttle lever. The smaller of the two controls idle speed, and the heavier spring regulates high speed. The heavy spring (Fig. 5-5) is mounted in an elongated slot at its upper end and does not come into play at low speeds.

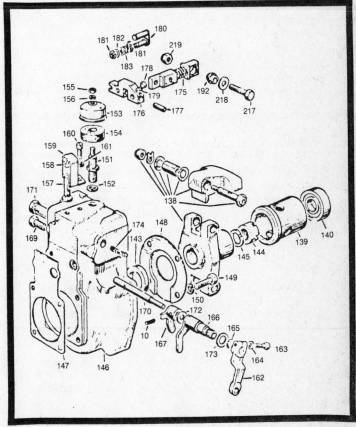

Fig. 5-5. Governor elements. (Courtesy GM Bedford Diesel.)

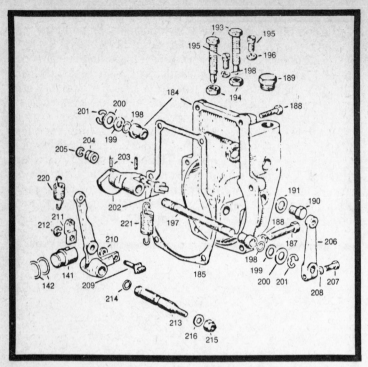

Fig. 5-6. Simms governor in exploded view.

As the engine starts, the fuel rack tends to cut off the fuel supply, and the engine slows. The governor weights generate less centrifugal force, and the springs then are able to pull the rack out toward the direction of more fuel. Operation is automatic, and the engine rpm is stable at whatever speed the operator has chosen.

The excess-fuel button opens the excess-fuel catch and allows the control lever to move forward beyond its normal limits during cold starts. Once the lever moves away from the catch, it locks; maximum speed is limited by stops on the lever.

**Dissembly**. This governor is, as stated earlier, simpler than most. But care should be taken to understand the function and interrelationship of the parts as they are disassembled Scrupulous standards of cleanliness should be maintained. Replace all gaskets and seals, and treat the springs with respect. Do not stretch or bend them. Parts numbers are keyed to Fig. 5-6.

Begin by undoing the bolt securing the control lever (*206*) to the shaft (*197*). Remove the lever. The governor rear casting is fastened to the front casting by six bolts. Separate the two castings enough to remove the Nyloc nut (219) and bolt (217) which secures the lever (209). The rear casting is free and can be set aside.

Remove the sleeve (139) with its bearing and for (141) from the weight assembly. Remove the control rod link from the control rod. It is held by a screw (176). Then remove the pin (168) from the maximum-fuel stop lever. Now take out the bearing (171) and shaft (170). Remove the spring clip (163) and stop assembly (166). The stop lever is now free.

The weight assembly must be removed with special tools. The Hartridge No. 87744 is needed to loosen the retaining nut (144), and the Hartridge puller No. 7044-48 to remove the weights.

The front half of the mechanism is held by four bolts. To disassemble further, remove the lower spring plate (211). It is held by an E-clip. Observe the lay of the springs for future reference; it is possible to assemble them wrong. Carefully remove the governor springs (221 and 220). Next, remove the Nyloc nut (215) and crank lever fulcrum shaft (213), along with the lever (209). Undo the full-speed and idle stops. The control lever cross-shaft is held by taper pins. Support the shaft and drive the pins out. Remove the arm (206) and the E-clips. Observe the number and location of any shims on the shaft.

**Assembly.** Insert the cross shaft (197) of the control lever in the governor rear half and assemble the stop control (206) and spring arm (202). Replace the taper pins (203), E-clips, and shims. If you have rebushed or replaced the cross-shaft, the end float should be adjusted by adding or substracting shims to between 0.05 and 0.25 mm. Thread the high-speed and idle stop screws into their bosses. The control stop can be fitted backwards if you are not observant.

Thread the fulcrum shaft (213) of the crank lever into the casting and assemble the crank lever (209) together with it. Use a new Nyloc nut (215) on the shaft. Place the governor springs in position and locate the plates on the small spring with E-clips.

Turn to the front casting. Fit the stop lever (166) and locate with a washer (173) and spring clip (165). The excess-fuel shaft (170) goes on the opposite side of the housing, along with the excess-fuel shaft (170), maximum stop lever (167), and washer (172). The shoulder of the shaft should extend through the stop lever. Secure the shaft with its pin and replace the spring (174) and the bearing (171). The bearing should be torqued to 16 ft-lb. The front casting is assembled with a new gasket. Note that the oil baffle (144) is part of the

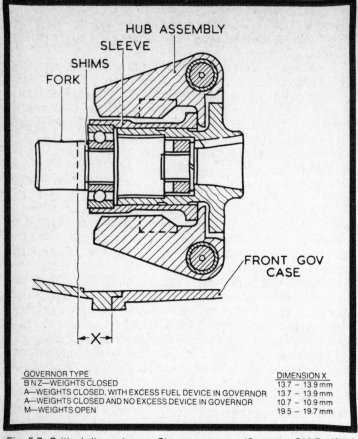

| GOVERNOR TYPE | DIMENSION X |
|---|---|
| B N Z—WEIGHTS CLOSED | 13.7 − 13.9 mm |
| A—WEIGHTS CLOSED, WITH EXCESS FUEL DEVICE IN GOVERNOR | 13.7 − 13.9 mm |
| A—WEIGHTS CLOSED AND NO EXCESS DEVICE IN GOVERNOR | 10.7 − 10.9 mm |
| M—WEIGHTS OPEN | 19.5 − 19.7 mm |

Fig. 5-7. Critical dimension on Simms governor. (Courtesy GM Bedford Diesel.)

package; torque to 6 ft-lb. Secure the link to the control rod and tighten.

To replace the weights you will need the Hartridge No. 87744 again. Replace the hub. If parts—hub, bearing, sleeve, fork—have been replaced check the dimension at $x$ in Fig. 5-7 and shim accordingly.

The telescopic link connects to the crank lever in the rear casting. Torque the castings together with the six bolts to 6 ft-lb. Replace the control lever (206) on the lever shaft (197).

Engine speed is a function of pump delivery as much as governor adjustment. After you are satisfied that the pump is delivering as it should, you can adjust the idle and high-speed stops to the engine. But if these stops are factory sealed they should not be disturbed.

## RSV Governor

The RSV governor requires some explanation, although it employs the familiar principle of flyweights acting in opposition to spring tension. Figure 5-8 shows the basic arrangement of the parts. The fuel rack (1) extends to the plungers and determines delivery per stroke. The floating lever (3) is suspended, pendulum fashion, at the pivot (B) and moves the fuel rack through the connecting link (2). The guide lever (4) is adjusted by the operator to determine speed. The shifter (5), which other manufacturers would describe as the *sleeve*, converts the angular motion of the weights (6) to fore and-aft movement. The weights are driven by the camshaft (7).

As speed increases, point A moves to the left in response to the displacement of the weights away from the camshaft. The shifter moves the lower end of the floating lever (A) to the left, and in so doing, moves the shared fulcrum (B) to the left. This movement causes the floating lever to retract the rack, reducing fuel delivery. Under load the mechanism moves to the right, reversing the procedure.

The drawing at Fig. 5-8 should be consulted as we get further into the discussion, since it is easy to lose sight of the fundamental principles involved under different speed and load conditions.

The next series of drawings are more realistic since they include all of the important functional parts. Be patient and

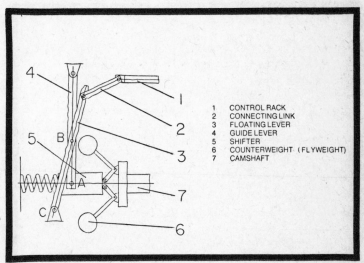

| 1 | CONTROL RACK |
| 2 | CONNECTING LINK |
| 3 | FLOATING LEVER |
| 4 | GUIDE LEVER |
| 5 | SHIFTER |
| 6 | COUNTERWEIGHT (FLYWEIGHT) |
| 7 | CAMSHAFT |

Fig. 5-8. Simplified drawing of the RSV governor. (Courtesy Marine Engine Div., Chrysler Corp.)

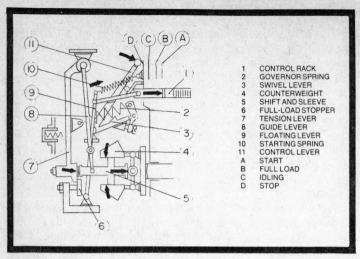

| | |
|---|---|
| 1 | CONTROL RACK |
| 2 | GOVERNOR SPRING |
| 3 | SWIVEL LEVER |
| 4 | COUNTERWEIGHT |
| 5 | SHIFT AND SLEEVE |
| 6 | FULL-LOAD STOPPER |
| 7 | TENSION LEVER |
| 8 | GUIDE LEVER |
| 9 | FLOATING LEVER |
| 10 | STARTING SPRING |
| 11 | CONTROL LEVER |
| A | START |
| B | FULL LOAD |
| C | IDLING |
| D | STOP |

Fig. 5-9. Start position. (Courtesy Marine Engine Div., Chrysler Corp.)

bear with me in this long and somewhat tedious explanation. Unless you understand how the mechanism works—either from the drawings or in conjunction with an actual governor—repair attempts are almost hopeless.

Figure 5-9 shows the governor at "start." The control lever (*11*) is set to the excess-fuel position; it moves the swivel lever to the right and stretches the governor spring (*2*). One

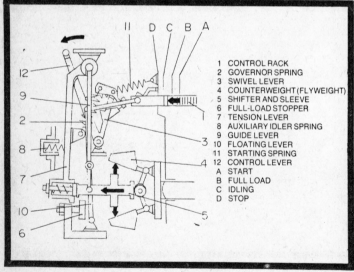

| | |
|---|---|
| 1 | CONTROL RACK |
| 2 | GOVERNOR SPRING |
| 3 | SWIVEL LEVER |
| 4 | COUNTERWEIGHT (FLYWEIGHT) |
| 5 | SHIFTER AND SLEEVE |
| 6 | FULL-LOAD STOPPER |
| 7 | TENSION LEVER |
| 8 | AUXILIARY IDLER SPRING |
| 9 | GUIDE LEVER |
| 10 | FLOATING LEVER |
| 11 | STARTING SPRING |
| 12 | CONTROL LEVER |
| A | START |
| B | FULL LOAD |
| C | IDLING |
| D | STOP |

Fig. 5-10. Idle position. (Courtesy Marine Engine Div., Chrysler Corp.)

end of the spring is connected to the tension lever (7) and brings the high-speed stop. When the engine starts, the weights are forced outward, overcoming the weak tension of the starting spring (10), and the shifter retracts until it comes into contact with the tension lever.

During idle the governor spring is slack (Fig. 5-10). It exerts little force on the tension lever. The weights are free to move outward, even at low rpm, and cause the tension lever to contact the auxiliary idler spring (8). The floating lever is moved to the idle position, severely restricting fuel delivery.

At maximum rpm the governor spring pulls the tension lever (Fig. 5-11) to the high-speed stop and moves the rack to full power. Should the engine speed fall off from this rpm, force on the weights will decrease, and the governor spring will cause the rack to delive more fuel.

Other features of this governor include an *angleich* (balance) spring as an override to provide more power at low rpm. The stop lever can be actuated at any speed, regardless of the position of the weights or other levers.

**Disassembly.** Using Fig. 5-12 as a guide, begin disassembly by separating the governor from the injector pump. Remove the *angleich* spring (32) and the auxiliary idler spring (58). Remove the lockscrew (51) from the housing. Raise the swivel lever (13) and lift out the tension lever (27) and spring (14). Now remove the sleeve (21) and guide lever (25). With the swivel lever raised, remove the floating lever (28) from the guide lever.

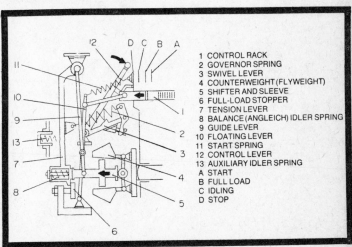

1 CONTROL RACK
2 GOVERNOR SPRING
3 SWIVEL LEVER
4 COUNTERWEIGHT (FLYWEIGHT)
5 SHIFTER AND SLEEVE
6 FULL-LOAD STOPPER
7 TENSION LEVER
8 BALANCE (ANGLEICH) IDLER SPRING
9 GUIDE LEVER
10 FLOATING LEVER
11 START SPRING
12 CONTROL LEVER
13 AUXILIARY IDLER SPRING
A START
B FULL LOAD
C IDLING
D STOP

Fig. 5-11. High-speed position. (Courtesy Marine Engine Div., Chrysler Corp.)

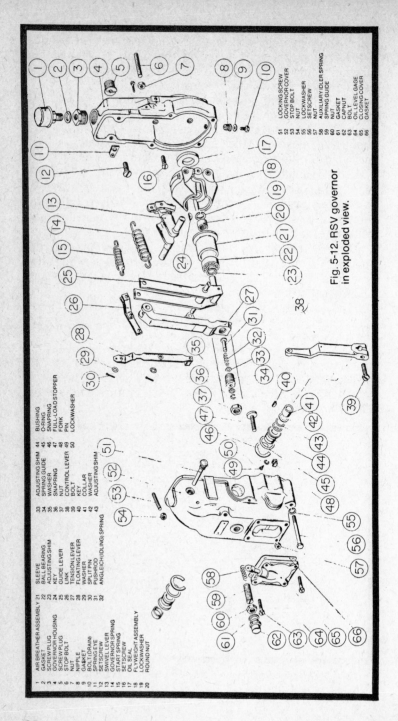

Fig. 5-12. RSV governor in exploded view.

| | | | | |
|---|---|---|---|---|
| 1 | AIR BREATHER ASSEMBLY | 21 | SLEEVE | |
| 2 | GASKET | 22 | BALL BEARING | |
| 3 | SCREW PLUG | 23 | ADJUSTING SHIM | |
| 4 | GOVERNOR HOUSING | 24 | KEY | |
| 5 | SCREW PLUG | 25 | GUIDE LEVER | |
| 6 | STOP BOLT | 26 | LINK | |
| 7 | NUT | 27 | TENSION LEVER | |
| 8 | NIPPLE | 28 | FLOATING LEVER | |
| 9 | GASKET | 29 | WASHER | |
| 10 | BOLT (DRAIN) | 30 | SPRING EYE | |
| 11 | SPRING EYE | 31 | SPLIT PIN | |
| 12 | SWIVEL LEVER | 32 | PUSHROD | |
| 13 | GOVERNOR SPRING | | ANGLEICH (IDLING) SPRING | |
| 14 | START SPRING | | | |
| 15 | SETSCREW | | | |
| 16 | SETSCREW | | | |
| 17 | OIL SEAL | | | |
| 18 | FLYWEIGHT ASSEMBLY | | | |
| 19 | LOCKWASHER | | | |
| 20 | ROUND NUT | | | |

| | | | | |
|---|---|---|---|---|
| 33 | SLEEVE | 44 | ADJUSTING SHIM | |
| 34 | BALL BEARING | 45 | SPRING GUIDE | |
| 35 | ADJUSTING SHIM | 46 | WASHER | |
| 36 | KEY | 47 | SNAPRING | |
| 37 | GUIDE LEVER | 48 | NUT | |
| 38 | LINK | 49 | CONTROL LEVER | |
| 39 | TENSION LEVER | 50 | BOLT | |
| 40 | WASHER | | | |
| 41 | FLOATING LEVER | | | |
| 42 | KEY | | | |
| 43 | ADJUSTING SHIM | | | |

| | | | |
|---|---|---|---|
| | BUSHING | | |
| | O-RING | | |
| | SNAPRING | | |
| 47 | FULL-LOAD STOPPER | | |
| | FORK | | |
| | PIN | | |
| | LOCKWASHER | | |

| | |
|---|---|
| 51 | LOCKING SCREW |
| 52 | GOVERNOR COVER |
| 53 | STOP BOLT |
| 54 | NUT |
| 55 | LOCKWASHER |
| 56 | SETSCREW |
| 57 | NUT |
| 58 | AUXILIARY IDLER SPRING |
| 59 | SPRING GUIDE |
| 60 | NUT |
| 61 | GASKET |
| 62 | CAP NUT |
| 63 | BOLT |
| 64 | OIL-LEVEL GAGE |
| 65 | CLOSING COVER |
| 66 | GASKET |

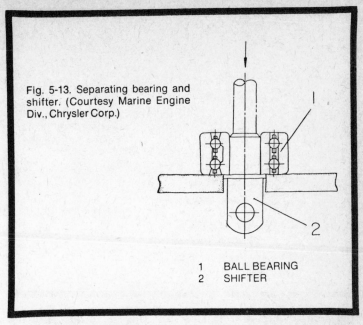

Fig. 5-13. Separating bearing and shifter. (Courtesy Marine Engine Div., Chrysler Corp.)

1   BALL BEARING
2   SHIFTER

If required, the bearing (22) can be pressed from the sleeve (21); support as shown in Fig. 5-13. Be extremely careful not to deface the surface of the shifter in the press operation. Next loosen the lockbolt and dismantle the control lever (38), along with assorted washers and shims (42 and 43). Do not remove the swivel lever (13) unless absolutely necessary. It is secured by snaprings. For governors so equipped, remove the lever (secured by a bolt on the lever shaft). The next step is to remove the end cover (65) and *angleich* spring assembly (31, 32, 32, and 34). It is held by a snapring.

Turn the bearing (22) by hand. It should turn freely and without undue noise. Check the swivel lever (13) bushings for excessive clearance; the lever should turn freely. Do the same for the pin bushing clearance at the weights (18); replace as needed. The spring should be tested by machine to determine if it has lost tension in service. Clean all parts thoroughly and lubricate prior to assembly.

**Assembly.** If a new bearing (22) has been installed, the sleeve (21) will have to be shimmed. Your dealer has the parts and instructions. The lockscrew (51) is torqued to 8 ft-lb. The distance from the push rod (31) to the tension lever (27) must be 1.5 mm, or 0.0587 in., as shown in Fig. 5-14. Use new gaskets and O-rings.

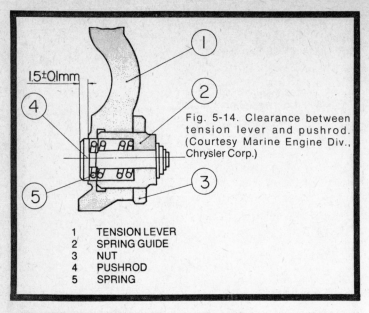

1.5±01mm

Fig. 5-14. Clearance between tension lever and pushrod. (Courtesy Marine Engine Div., Chrysler Corp.)

1   TENSION LEVER
2   SPRING GUIDE
3   NUT
4   PUSHROD
5   SPRING

**Adjustment.** Adjustments for this governor are not simple—at least, not simple if you want the performance and responsiveness which this device is capable of. You will need a test bench with a variable-speed drive and dial indicator to determine rack movement as a function of rpm. With the injector pump on the machine, match the indicator with zero extension of rack. Operate the control lever and determine that full stroke is 21.0 mm, or 0.827 in. The rack should move smoothly and should be impelled to maximum fuel injection by the force of the starting spring. The stop bolt bearing against the control lever should be set to correspond with the 0.5−1.0 mm (0.0197−0.039 in.) control rack position.

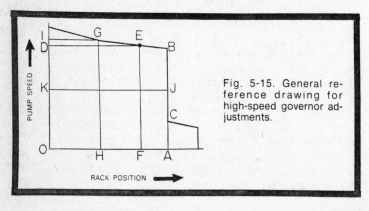

Fig. 5-15. General reference drawing for high-speed governor adjustments.

Begin with the high-speed adjustment. Remove the end cover. You will see the full-load stop screw (Fig. 5-12, No. 47) at the lower part of the unit. Loosen its locknut and turn the screw so that the rack position corresponds to A in Fig. 5-15 at pump speeds between B and C. To increase the fuel delivery—i.e., lengthen the inward movement of the rack—turn the screw clockwise. The maximum speed is represented by point G. Adjust the stop screw (Fig. 5-12, No. 6) to obtain the requisite rack position at this rpm.

Next we have the matter of governor sensitivity. Referring again to Fig. 5-15, G represents maximum no-load speed. Point E represents the rated speed. The difference between E and G is expressed as a percentage. Subtract the rated speed from the maximum no-load speed and divide the remainder by the rated speed. For example, if the rated speed were 3200 rpm and no-load speed reached 3400, we would divide the difference, 200 rpm, by 3200 and multiply by 100 to convert the answer to a percentage. Variation would be 6.25%, which is quite acceptable for most applications.

If the variation is beyond specs, remove screw plug (Fig. 5-12, No. 3) from the top of the governor casting and turn the adjusting knuckle screw. Tightening the screw reduces the percentage of variation; loosening it increases the variation. At the same time, the position of the screw influences rated speed by changing governor spring tension (Fig. 5-16). Rated speed must be reset after this adjustment.

> *Caution*: Do not unscrew the adjustment more than 20 clicks (4 clicks to the turn).

Most RSV governors include an angleich spring assembly. The pump-rpm, rack-position curve looks like the one in Fig. 5-17. The angleich spring is responsible for the dogleg from B to C. Set the control lever to maximum speed and operate at F

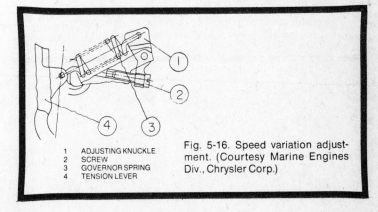

| 1 | ADJUSTING KNUCKLE |
| 2 | SCREW |
| 3 | GOVERNOR SPRING |
| 4 | TENSION LEVER |

Fig. 5-16. Speed variation adjustment. (Courtesy Marine Engines Div., Chrysler Corp.)

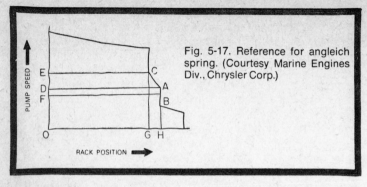

Fig. 5-17. Reference for angleich spring. (Courtesy Marine Engines Div., Chrysler Corp.)

rpm. Tighten the angleich spring assembly with the appropriate wrench. You can fabricate one in the shop or purchase Chrysler's tool number 47916-212. Tighten until the control rack moves to *H*. Tighten the locknut to hold the adjustment.

To check the operation of the spring, mount the pump assembly on the engine. Slowly accelerate from speed *D*. At rpm *E* the rack should move from *H* to *G*. Simple?

The idle adjustment is made at the after end of the governor. The idler spring guide, shown as *59* in Fig. 5-12, can be threaded in and out to adjust the tension on the idle spring. To adjust, set the control lever to the stop position, so that the control rack is at *B*. Set the pump speed to *D* and adjust spring tension to move the rack to *C*. Do not overtighten the spring. The idle adjustment curve is shown in Fig. 5-18.

The rack-position, pump-rpm relationship varies between engine makes, models, and to some extent, governor model. No attempt has been made here to include this information, since it is available from dealers and diesel pump specialists.

## PNEUMATIC GOVERNORS

Pneumatic, or *flap valve*, governors are generally less expensive than the more sophisticated centrifugal types, but

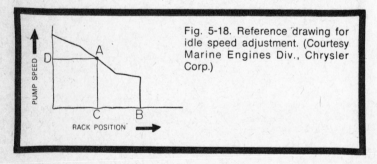

Fig. 5-18. Reference drawing for idle speed adjustment. (Courtesy Marine Engines Div., Chrysler Corp.)

are not as progressive in their action (although some engineers would argue that the difference is academic).

These governors operate on the *venturi* principle. When a moving fluid encounters a restriction, its velocity increases. And, in accordance with the principle that we cannot get something for nothing, its pressure falls. Figure 5-19 illustrates the classical venturi—a smoothly narrowing restriction in a bore. The numbers are arbitrary and could express velocity in feet per second, miles per hour, or any other measure. Pressure could be expressed as psi, atmospheres, or inches of mercury. What is interesting about the venturi is that each one has a *constant*. In this case, the constant is 20. If we multiply pressure times velocity at any point in the pipe the answer is always the same. Of course, in the real world, things are not so neat, and our constant tends to be fuzzed by turbulence, air resistance, and other factors.

The vacuum the venturi draws is a function of air velocity through it. The faster the column of air moves, the more vacuum, or negative pressure. Since air velocity in a diesel engine is dependent upon piston speed, we can use venturi-induced vacuum to monitor rpm. The venturi need not be a streamlined restriction such as the one shown in Fig. 5-19; any restriction, such as provided by the edge of the flap valve, will do.

Vacuum is conveyed, by means of a tube, from the venturi to a diaphragmed chamber at the governor. The diaphragm is spring-loaded and free to move fore and aft in the housing. The movement is transferred to the fuel by means of a link.

### MZ Pneumatic Governor

Refer to Fig. 5-20 for a structural view of this device. The air intake forms the primary venturi (2), with a secondary

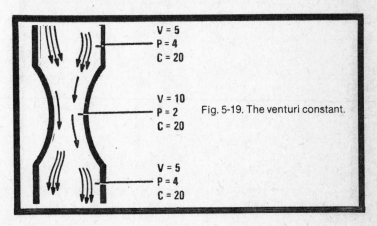

Fig. 5-19. The venturi constant.

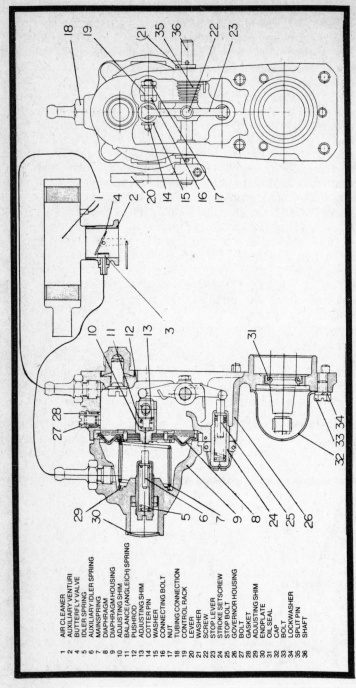

Fig. 5-20. The MZ governor. (Courtesy Marine Engines Div., Chrysler Corp.)

| | |
|---|---|
| 1 | AIR CLEANER |
| 2 | AUXILIARY VENTURI |
| 4 | BUTTERFLY VALVE |
| 5 | IDLER SPRING |
| 6 | AUXILIARY IDLER SPRING |
| 7 | MAINSPRING |
| 8 | DIAPHRAGM |
| 9 | DIAPHRAGM HOUSING |
| 10 | ADJUSTING SHIM |
| 11 | BALANCE (ANGLEICH) SPRING |
| 12 | PUSHROD |
| 13 | ADJUSTING SHIM |
| 14 | COTTER PIN |
| 15 | WASHER |
| 16 | CONNECTING BOLT |
| 17 | NUT |
| 18 | TUBING CONNECTION |
| 19 | CONTROL RACK |
| 20 | LEVER |
| 21 | WASHER |
| 22 | SCREW |
| 23 | STOP LEVER |
| 24 | STROKE SETSCREW |
| 25 | STOP BOLT |
| 26 | GOVERNOR HOUSING |
| 27 | BOLT |
| 28 | GASKET |
| 29 | ADJUSTING SHIM |
| 30 | ENDPLATE |
| 31 | OIL SEAL |
| 32 | CAP |
| 33 | BOLT |
| 34 | LOCKWASHER |
| 35 | SPLIT PIN |
| 36 | SHAFT |

near the flap or *butterfly* valve edge (3). A tube bleeds vacuum from these venturis to the left, or low-pressure, side of the diaphragm housing. A second tube brings filtered air to the atmospheric side of the chamber. The diaphragm (8) separates these two halves of the housing. It is loaded by the mainspring (7) and connected to the fuel rack.

As long as the flap valve remains stationary in the air intake bore, vacuum in the left side of the chamber is constant. The diaphragm takes a position determined by the spring and the pressure differential on either side of it. Should the engine speed up, vacuum will increase in the left chamber, and the diaphragm, impelled by atmospheric pressure on the right, moves to the left. The mainspring is compressed and the rack is pulled out, reducing fuel delivery. If the engine decelerates, as under load, vacuum decreases and the diaphragm shifts to the right under spring tension. The rack moves in and more fuel is provided. Again the system stabilizes at the predetermined speed fixed by the angle of the flap valve in the air intake.

At full power the valve is open, causing less vacuum to be generated in the venturi, with a consequent increase of fuel delivery as the mainspring pushes the diaphragm to the right. At intermediate valve positions there is correspondingly less fuel delivery. There is no direct connection between the throttle and pump; all signals are delivered by means of pressure differentials in the diaphragm chamber.

The stop lever (23) has two adjustments: The stop bolt (25) limits the maximum fuel delivery during cold cranking, and the stroke setscrew (24) limits fuel once the engine starts. The fuel enrichment provision is by means of a spring which temporarily overrides the stroke setscrew adjustment.

Figure 5-21 shows this relationship graphically. If maximum injection is set at A for best low-speed running, it will increase to B' at power; and if we adjust output for best high speed (B), the engine will be starved at low speed, as indicated by A'. Ideally the mechanism should be set to A and progress to B at high speed.

The *angleich* mechanism is shown in detail in Fig. 5-22. It acts to reduce fuel delivery at low rpm by moving the fuel rack farther out than would otherwise be the case. At low speeds vacuum in the right chamber is low; the diaphragm moves to the left, impelled by the mainspring. As it does it compresses the *angleich* spring inserted behind the push rod (Fig. 5-22, No. 9) and allows the rod to move nearer the zero-delivery position. As the engine accelerates, the diaphragm moves to the right. The balance spring mechanism moves with it, away

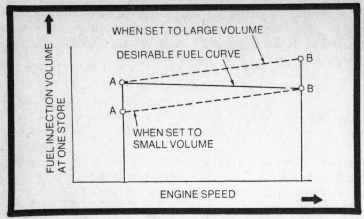

Fig. 5-21. Fuel injection curves. (Courtesy Marine Engine Div., Chrysler Corp.)

from the stop lever. Once it is clear of the lever, the spring has nothing to react against (except the plungers, which should rotate with a force of a few grams) and is out of the circuit.

**Disassembly.** Referring to Fig. 5-20, remove the governor from the pump at the diaphargm chamber mounting screws, and disconnect the vacuum and air lines and the fuel rack link. Begin disassembly of the governor proper by unscrewing the idler spring assembly (5) from the housing. Remove the push rod (12), adjusting shim (13), and balance spring (11) from the diaphragm. Next remove the stroke setscrew (24), camshaft cap, and the stop lever (23), along with its mounting hardware.

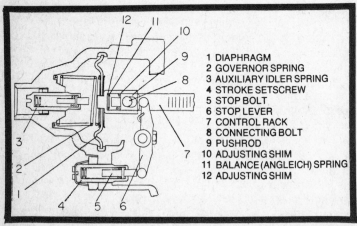

1 DIAPHRAGM
2 GOVERNOR SPRING
3 AUXILIARY IDLER SPRING
4 STROKE SETSCREW
5 STOP BOLT
6 STOP LEVER
7 CONTROL RACK
8 CONNECTING BOLT
9 PUSHROD
10 ADJUSTING SHIM
11 BALANCE (ANGLEICH) SPRING
12 ADJUSTING SHIM

Fig. 5-22. Balance mechanism. (Courtesy Marine Engine Div., Chrysler Corp.)

Remove the stop lever (23) and, after extracting the cotter pins from either end, remove the shaft (36). Finally, remove the diaphragm and spring.

Clean all parts thoroughly, with special attention to the diaphragm. It is made of leather and can become stiff in service. Holding it by the outer ring, pull upward on the center section. It should collapse of its own weight. If it does not, apply diaphragm oil (available from your distributor) to soften the leather or, in doubtful cases, replace the diaphragm. The spring should be inspected for wear and its free length compared to a new one. For absolute reliability you should check the spring tension against the manufacturer's specs. Examine the top lever shaft for excessive clearance. You may have to replace both the shaft and governor housing.

**Assembly.** Assemble in the reverse order. The *angleich* spring is shimmed as shown in Fig. 5-23 to adjust the stroke of the push rod. Specifications vary with the application, but for the Chrysler Nissan series they are as follows:

SD22: 1 mm ( 0.039 in.)
SD33: 0.6 mm (0.0236 in.)

When installing the stroke setscrew (24), make sure that it contacts the end of the stop lever.

**Adjustment.** Mount the governor pump assembly on the test machine. Set the rack and indicator at zero and connect a vacuum line to the left diaphragm chamber. Run the pump at 500 rpm during the test. (The turning resistance of the plungers varies with speed; static tests are almost meaningless.)

Apply a vacuum of 500 mm of water to the diaphragm and verify that the rate of drop averages no more than 2 mm/sec

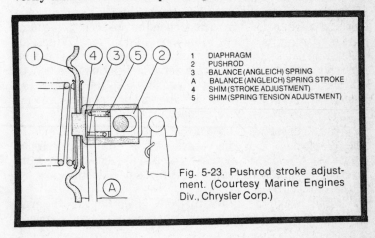

1   DIAPHRAGM
2   PUSHROD
3   BALANCE (ANGLEICH) SPRING
A   BALANCE (ANGLEICH) SPRING STROKE
4   SHIM (STROKE ADJUSTMENT)
5   SHIM (SPRING TENSION ADJUSTMENT)

Fig. 5-23. Pushrod stroke adjustment. (Courtesy Marine Engines Div., Chrysler Corp.)

for 10 sec. Adjust the stroke setscrew so that the control rack is on specification under full vacuum. If the stroke of the rack is less than prescribed, loosen the screw; if more, tighten it. Check the adjustment of the *angleich* spring again, as shown in Fig. 5-23. Some discretion is required here since the specifications can be voided if the spring resists movement. The best advice is to compare the action of the mechanism with a known-good assembly, and shim for tension as well as stroke.

The mainspring controls the movement of the fuel rack, and adjustment is quite critical. Compare the output with the manufacturer's curve. Increasing the number or thickness of the shims will compress the spring and increase starting pressure.

The idling adjustment has implications over the whole operating range. If the control rack movement does not correspond with the desired curve, you may assume that the idle speed is too high or too low. Tightening the auxiliary idle spring increases the speed. You should use a factory tool when making this adjustment (available from Chrysler as No. 57915-422).

**Venturi.** The venturi section seldom needs attention. After long use the shaft (Fig. 5-24, No. *16*) may wear and need replacement. Disassemble using the drawing as a guide. On assembly, shim the shaft so that the flap valve (*17*) moves its full range without binding against the venturi casting.

### Simms GP Pneumatic Governor

The Simms unit is similar to the MZ, less the *angleich* spring (see Fig. 5-25). The throttle valve A is connected to the throttle lever or pedal and is the only means of speed control. The governor mounts on the back of the injection pump and is spring-loaded to hold the rack open in the absence of vacuum. The diaphragm is similar to the one used on the MV design, consisting of a leather cup secured at the edges by the housing castings. The link assembly transfers diaphragm movement to the fuel rack. The damping valve, shown to the right of the governor unit, gives smooth operation by damping diaphragm surge. It is secured by an external locknut.

The mainspring is sandwiched between the right side of the diaphragm and the inner wall of the low-pressure chamber. A stop lever moves the rack to the zero-fuel position, overriding the governor. An excess-fuel device is fitted to most models for cold starting.

Adjustment is by turning the valve guide J to vary the area of the inlet port G. Air at near atmospheric pressure (minus

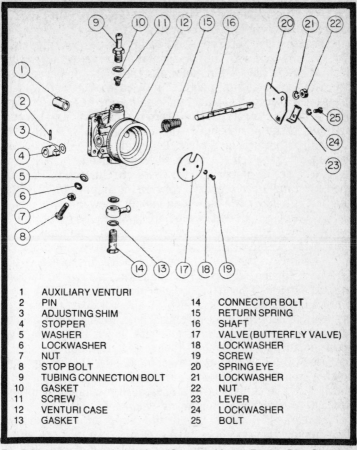

| | | | |
|---|---|---|---|
| 1 | AUXILIARY VENTURI | | |
| 2 | PIN | 14 | CONNECTOR BOLT |
| 3 | ADJUSTING SHIM | 15 | RETURN SPRING |
| 4 | STOPPER | 16 | SHAFT |
| 5 | WASHER | 17 | VALVE (BUTTERFLY VALVE) |
| 6 | LOCKWASHER | 18 | LOCKWASHER |
| 7 | NUT | 19 | SCREW |
| 8 | STOP BOLT | 20 | SPRING EYE |
| 9 | TUBING CONNECTION BOLT | 21 | LOCKWASHER |
| 10 | GASKET | 22 | NUT |
| 11 | SCREW | 23 | LEVER |
| 12 | VENTURI CASE | 24 | LOCKWASHER |
| 13 | GASKET | 25 | BOLT |

Fig. 5-24. Venturi in exploded view. (Courtesy Marine Engine Div., Chrysler Corp.)

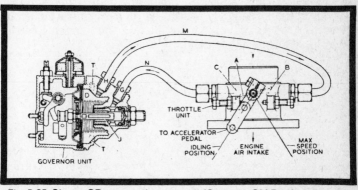

Fig. 5-25. Simms GP pneumatic governor. (Courtesy GM Bedford Diesel.)

97

the drop across the filter element) enters through the tube $N$. The valve is, in effect, a buffer spring.

**Disassembly.** Separate the housing halves by removing the Phillips screws. The diaphragm can be removed by unhooking the link at the diaphragm rod. The diaphragm is delicate and should be handled with care. You will find a tapered lockscrew on the underside of the governor. It secures the link to the control rod. Next, remove the splined stop lever by opening the pinchbolt. If you wish to dismantle the unit further, withdraw the excess-fuel shaft by removing the lockpin and the bearing, which is screwed into place.

Examine the diaphragm for tears and surface cracks. It should be pliable. To replace it remove the diaphragm from its shaft (it is held by two nuts) and install a new one which has been soaked for at least one-half hour in Shell grade C calibration fluid. The filter element should be soaked in solvent and wetted with heavy oil.

**Assembly.** Assembly is straightforward and presents no serious problems. The governor mounting screws should be tightened to 5 ft-lb. Specifications are not given for the Phillips screws, but they should be drawn up in a crisscross manner to prevent the housing from warping. The damping valve should be coated with a colloidal graphite lubricant. *Oildag* is recommended—but you might have to go to England to get it.

Check the unit for leaks by compressing the spring and holding your fingers over the air tube fittings. The diaphragm should move and then stop as air pressure equals spring force. Continued movement means an air leak.

**Adjustment.** First, calibrate the pump on a test bench. Then mount the assembly on the engine; after a few minutes of warmup. set the maximum no-load speed at the throttle stop. Next, set the idle stop screw to obtain the specified rpm and adjust the damping valve guide $J$ for smoothest idle. The guide is secured by a locknut which can allow air to leak into the chamber if backed out too far. Keep it finger tight during adjustment.

# Cylinder Heads and Valves

**6**

A vast amount of developmental work was done in the 1920s and 1930s on the optimal shape of the combustion chamber. The work continues, but the main outlines of the problem and the alternatives appear to be firmly established.

For fuel to burn it must be heated above the ignition point in the presence of free oxygen. The fuel spray must be scattered about the cylinder so that all oxygen can take part in combustion, and should be atomized as finely as consistent with penetration. As the particles fan away from the nozzle, the chances for complete combustion are progressively lessened. There is less oxygen available since some of it has already combined with fuel droplets, and the size of the individual droplets shrinks as successive shrouds of vapor boil off and ignite.

The odds for complete combustion are improved by introducing some degree of controlled turbulence. The injectors have all they can do to atomize the charge and can at best only give a slow spin to the spray pattern. The alternative is to agitate the oxygen-bearing air around the fuel. This is done by restrictions or exitways angled in at the intake ports and valve seats, or, more popularly, by means of combustion chamber shape.

## COMBUSTION CHAMBERS

Diesel engine combustion chamber shapes are divided into two groupings, with some overlap. Open chambers resemble those employed in spark ignition (SI) engines: There is a single chamber above the piston crown where combustion takes place (Fig. 6-1). The piston or the head may be recessed to conform to the fuel spray pattern and achieve the desired compression ratio. The edges of the piston crown and lower

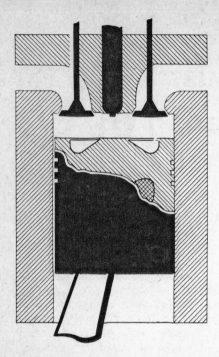

Fig. 6-1. Open chamber.

roof of the chamber are parallel, to form a "squish" area. Air
is trapped between these two faces and rushes out toward the
center of the chamber as the piston approaches TDC. The spray
pattern is typically flat, at an included angle of 160° or more to
insure good penetration. This direct injection chamber
requires, in general, a finer mist than the more elaborate
types. The advantages include a low volume ratio, which
means less heat lost to the cylinder walls and low pumping
losses.

Divided chambers are in several styles. The
*precombustion chamber* (Fig. 6-2) was the original chamber
and actually predates the diesel engine. The low-compression
Hornsby-Ackroyd oil engine used such an antechamber. As
shown in Fig. 6-2, the smaller chamber, accounting for
25—40% of the main chamber in volume, is connected to the
main chamber by a narrow passage. During the compression
stroke air is introduced into the precombustion chamber.

The fuel charge then is injected directly into it for ignition.
There is not enough oxygen in the chamber for complete
combustion, but enough is present to raise chamber pressures.
The unburned charge rushes out of the chamber and mixes
with air in the main space over the piston. The fuel is dispersed

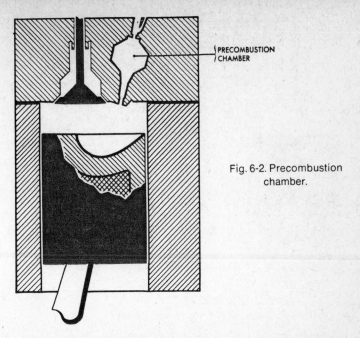

Fig. 6-2. Precombustion chamber.

by its own heat energy, rather than by virtue of high pump pressures and fine injector orifices. Consequently, injection gear is simplified and requires less maintenance. Additional benefits include better, or at least more consistent, fuel economy and the ability to burn a wide range of fuels. Because most of these chambers are uncooled, they are referred to as *hot bulb* chambers.

The *turbulence chamber* (Fig. 6-3) is similar in appearance to the precombustion chamber, but its function is different. Both the piston crown and chamber roof are flat, and they form a giant "swish" area. The chamber is usually spherical and is connected to the clearance volume (space above the piston by a tangential passageway. As the piston approaches TDC, the crown partially masks the entrance to the passageway. Consequently, the velocity of the air entering the chamber is increased. This turbulence speed is approximately 50 times crankshaft speed. Fuel injection is timed to occur at the velocity peak. The incoming fuel spray is exposed to fresh air for the duration of injection. The angle of the passageway and the shape of the turbulence chamber produce a swirl to the air and flaming fuel particles. As the piston passes TDC, the direction of movement reverses, and fuel and air swirl out the precombustion chamber and into the cylinder proper. Expansion takes place against the piston.

101

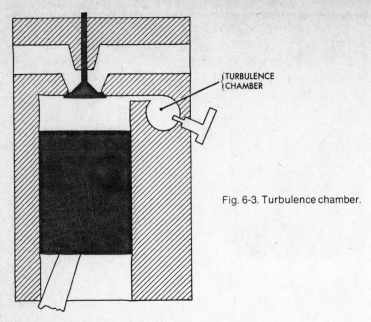

Fig. 6-3. Turbulence chamber.

The precombustion chamber has between a quarter and half of the clearance volume. Its function is to precondition the fuel for combustion, most of which takes place in the main chamber. The turbulence chamber accounts for almost all of the volume above the piston. Combustion occurs mainly in the chamber and not above the piston. The area above the piston does contain some oxygen, and combustion does continue in this space; but it may be best thought of as an arena for air/fuel mixing. The precombustion chamber is subject to a simple in-and-out movement of compressed air and expelled fuel. Movement in the turbulence chamber has a definite swirl, first in one direction and then, as combustion commences, in the other.

The third type of chamber is known under several names. The trade name is *Lanova chamber*, but is also called a *divided chamber* or an *energy cell*. This chamber used widely in automotive and tractor engines, has advantages which make the somewhat eleborate foundry work justified.

Figure 6-4 shows the figure-8 configuration of the chamber and the opposed injector and antechamber. As in the open chamber, the main volume of air remains above the piston. Most combustion takes place in this space. As in the turbulence chamber, the design generates a high degree of turbulence. But this system does not entail pumping losses (energy is required to force air through the narrow passage

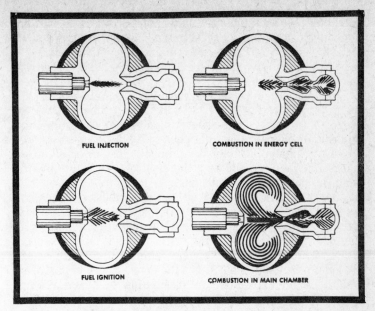

FUEL INJECTION

COMBUSTION IN ENERGY CELL

FUEL IGNITION

COMBUSTION IN MAIN CHAMBER

Fig. 6-4. Lanova divided chamber—fuel combustion.

leading to the turbulence chamber, which could otherwise be used to turn the crankshaft). Nor is turbulence a function of piston speed in the Lanova design. Turbulence is the result of thermal expansion and is independent of rpm.

The antechamber, or energy cell, consists of an inner and outer chamber. The inner chamber, which is the smaller of the two, opens to a narrow throat, situated between the lobes of the main combustion chamber. Readers familiar with SI engines will recognize the funnel-shaped throat as a venturi not unlike those employed in carburetors. The larger chamber communicates with the inner chamber through a second venturi.

During the compression stroke about 10% of the air volume passes into the energy cell, while the remainder stays above the piston. Fuel is injected in a solid stream which passes through the small axis of the combustion chamber and into the energy cell. Some small percentage of fuel shears off from the main stream and is ignited in the two lobes of the main chamber, but this is incidental. The bulk of the fuel is trapped in the small inner chamber of the energy cell. Some manages to traverse the second venturi and enter the outer chamber. There it encounters a mass of superheated air and explodes. Pressure rises violently, and the outer cell empties back through the venturi into the second cell and drives the unburned fuel back into the main chamber above the piston. The kidney-shaped walls of the main chamber impart a swirl to the fuel charge so that it mixes thoroughly with the air.

The venturis slow blowback to give a gradual pressure rise against the piston.

In addition to providing turbulence which is constant regardless of load or speed, the divided chamber allows the use of a simple, single-orifice injector nozzle. Studies by Caterpillar indicate that this chamber type produces lower emissions than the others, probably because of more consistent fuel mixing and the control exercised by the venturis on the rate of combustion.

Besides forming at least part of the combustion space, the cylinder head forms a major part of the cooling system, and it must be sturdy enough to contain the pressures generated in the chambers. The great majority of heads are cast iron, although a few air-cooled engines employ aluminum heads. Aluminum has about four times the conductivity of cast iron, which is an advantage both for dissipating the heat developed in combustion and for promoting a uniform heat gradient across the head.

The water jacket should be cast with large passages which will resist clogging. But merely pumping coolant through the head will not insure that local hot spots are cooled. Some designs employ replaceable diverters to direct the stream at the valve seats and other critical areas. Generally the head forms an integral circuit with the block. There is a recent tendency to do away with interconnected water passages and keep the block and head circuits separate. The advantage of this approach is that head gasket failure does not contaminate the oil with water. Contamination becomes quite serious when the water is mixed with ethylene glycol (antifreeze).

For the sake of savings in materials if not weight, the water jacket should form part of the head structure; it should bear part of the loads. The fasterners—either capscrews or, preferably, studs—should be arranged symmetrically in so far as the rocker pedestals, compression release, and other subordinate gear allow.

From these remarks you can appreciate that cylinder head design is one of the most demanding aspects of engine building. The general rule is to stick with old, proven designs and look with jaundiced eye at anything new, even if the computer says it will work.

## VALVES AND VALVE GEAR

While a few piston-ported 2-cycles survive, most of these engines employ exhaust valves. A few use valves in the inlet side, although this sort of complexity is usually limited to large engines. Ports serve as well and have the advantage of no additional moving parts. Because of the limited amount of time available for 2-cycle functions—the complete operating cycle takes place in one revolution of the crankshaft—2-cycle valves are often paired in the cylinder to increase their area. The Detroit Diesel models represent an example of this practice. Four-cycle engines must, of course, employ intake as well as exhaust valves. Some employ paired valves because of space restrictions in the head and to reduce the inertia of each valve for better control. The type of valve used (Fig. 6-5) is called a *poppet*, *mushroom*, or *tulip* valve.

Because of the temperatures involved, intake valves are made of chromium—nickel alloy and the exhaust valves of silicon and chromium. The intake valve has the benefit of cooling air and is less critical than the exhaust, which is "scrubbed" by the exhaust gases. Some exhaust valves are filled with sodium. When heated, sodium becomes a liquid and

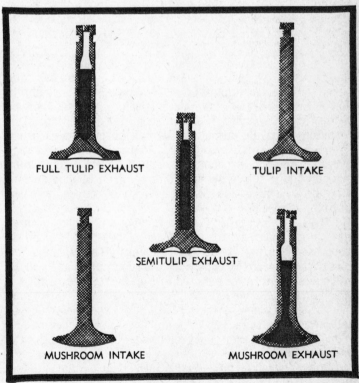

FULL TULIP EXHAUST

TULIP INTAKE

SEMITULIP EXHAUST

MUSHROOM INTAKE

MUSHROOM EXHAUST

Fig. 6-5. Valve styles—the solid shading represents sodium.

transfers heat to the valve guide. Sodium combines explosively with water, and these valves should be treated with respect. When discarded, they should be clearly identified as containing sodium.

The valve inserts, or seats, are a shrink interference fit in the cylinder head. In addition to forming a gas seal with the valve face, the seats provide the main path for heat transfer from the valve head. Consequently, the seats are adjacent to fins in air-cooled designs, and adjacent to the water jacket in the more popular liquid-cooled types. The bevel is self-centering and can correct some degree of valve guide wear. Normally, the intake seat is at an angle of 30° and the exhaust is at 45°. The 45° bevel gives a 20% higher seating load than the more shallow angle, but it exacts a penalty in the form of reduced flow. As a rule of thumb, a 45° valve must be lifted 20% more to obtain the same gas flow as a 30° valve with the same area. For this reason the intake is normally at 30°, and the exhaust, which operates in more severe environment, is ground to 45°. Usually the cam lift is the same for both, but exhaust restrictions are less meaningful since most of the gases escape during blowdown, at pressures considerably higher than atmospheric.

While the cam forces the valves open, only spring pressure closes them (Fig. 6-6). In some designs the cam has little effect on the upper limit of lift since acceleration is so abrupt that the inertia of the cam follower, push rod, valve, and part of the rocker arm drives the valve ahead of the cam. This condition is hardly desirable and can be cured by careful attention to the cam profile and to limiting reciprocating masses. Another problem associated with reciprocating motion occurs during seating. Many valves bounce several times before coming to rest. The severity of bounce increases with engine speed.

The valve guides are usually made of a phosphor-bronzed compound and pressed into place. They are second in importance to the seats as heatsinks. The springs are contained between two caps and secured by split collars on the valve stem. The valve pictured in Fig. 6-7 features a rotator which turns the valve a few degrees during each lift. Rotating valves is not absolutely essential; new processes, such as spraying the valve faces with aluminum, have almost made it redundant. But rotation does prolong seat life by averaging the wear over the whole surface.

*Valve lash* is the total of clearances between the cam and the cam follower (tappet), both ends of the push rod, and the rocker arm and valve stem. Some running clearance is

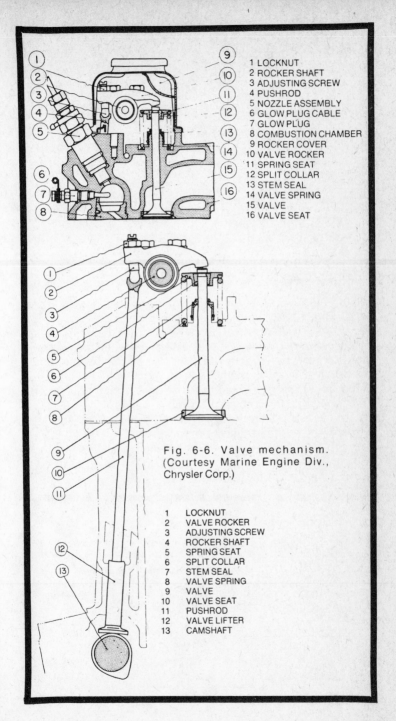

1 LOCKNUT
2 ROCKER SHAFT
3 ADJUSTING SCREW
4 PUSHROD
5 NOZZLE ASSEMBLY
6 GLOW PLUG CABLE
7 GLOW PLUG
8 COMBUSTION CHAMBER
9 ROCKER COVER
10 VALVE ROCKER
11 SPRING SEAT
12 SPLIT COLLAR
13 STEM SEAL
14 VALVE SPRING
15 VALVE
16 VALVE SEAT

Fig. 6-6. Valve mechanism. (Courtesy Marine Engine Div., Chrysler Corp.)

1    LOCKNUT
2    VALVE ROCKER
3    ADJUSTING SCREW
4    ROCKER SHAFT
5    SPRING SEAT
6    SPLIT COLLAR
7    STEM SEAL
8    VALVE SPRING
9    VALVE
10   VALVE SEAT
11   PUSHROD
12   VALVE LIFTER
13   CAMSHAFT

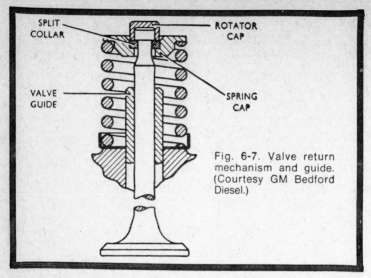

Fig. 6-7. Valve return mechanism and guide. (Courtesy GM Bedford Diesel.)

SPLIT COLLAR

ROTATOR CAP

VALVE GUIDE

SPRING CAP

required to insure that the valve will seat. Should it not make solid contact with the seat, compression will be lost and the valve will quickly overheat. Because of expansion most engine makers specify that clearances be adjusted when the engine is hot. However, as a practical matter, the reason for the rather liberal clearances on many engines is not to allow space for valve stem growth. Instead, it is the final timing adjustment. (Increasing the lash retards the timing; narrowing it opens the valves earlier.) Most designs are adjusted by means of a screw at the push rod side of the rocker arm, as shown in Fig. 6-6. The GM 2-cycle employs a threaded push rod which is rotated in or out to set the lash (Fig. 6-8), while other designs are adjusted from the rocker stud.

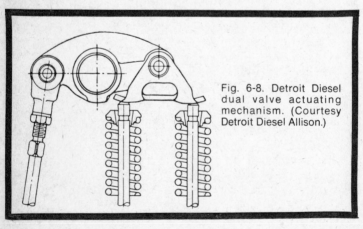

Fig. 6-8. Detroit Diesel dual valve actuating mechanism. (Courtesy Detroit Diesel Allison.)

# CYLINDER HEAD SERVICE

Lifting the cylinder head is a fairly serious operation and should not be entered into unless you know exactly what you are looking for. Most cylinder head difficulties are caused by leaks. Coolant may enter the oil sump by way of the head gasket, the injector tubes, or cracks in the casting. Compression may leak between adjacent cylinders to the atmosphere, or to the coolant and out through the radiator. In some instances the valve guides may be worn enough to allow lube oil to be sucked into the cylinder. However, when the guides are gone, the matter of removing the cylinder head becomes less significant since the whole engine will have to be torn down to replace rings, pistons, and so forth.

Cracks in the cylinder head can be caused by overtightening the head bolts or by tightening them in the wrong sequence (see Fig. 6-9). On some engines over-tightening the injector clamp bolts will have the same effect. Excessive fuel may raise combustion pressures beyond the strength of the head material. The fuel may originate at dribbling injectors, or can be pulled over from an oil-bath air cleaner which is overfilled. Scale in the water jacket or dirt deposits on the fins can also cause overheating and consequent head damage. A sixteenth of an inch of scale adhering to 1 in. of cast iron equals 4¼ in. of cast iron in insulating effect.

A weak cylinder can be located by disabling one injector at a time. A screwdriver blade may be used to hold down unit injectors, or the fuel line can be opened. Note the rpm drop as each cylinder is cut out. A cylinder which causes less of a drop than the others is defective. Exhaust temperature gages are very useful monitors, especially if senders for each cylinder are fitted (Fig. 6-10).

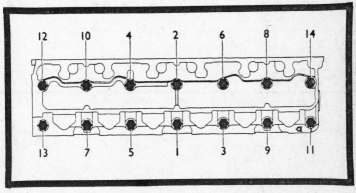

Fig. 6-9. Typical torquing sequence. (Courtesy GM Bedford Diesel.)

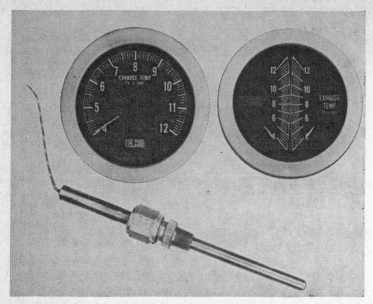

Fig. 6-10. Compressing valve spring. (Courtesy GM Bedford Diesel.)

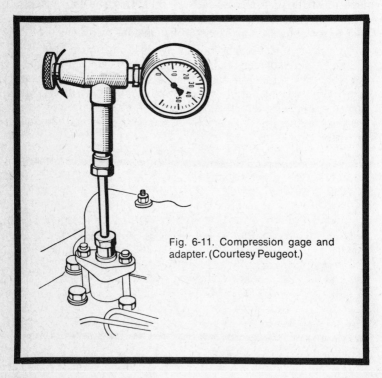

Fig. 6-11. Compression gage and adapter. (Courtesy Peugeot.)

A compression check is also useful, especially when the check is part of an on-going maintenance program. Individual checks can mean little since so many variables can intervene in a given test. The gage (Fig. 6-11) is inserted at the injector boss by means of an adapter. Some manufacturers specify that the engine be at operating temperature, while others prefer cranking readings. A cranking test is less reliable since the ambient air temperature, oil viscosity, battery state of charge, and starter circuit are variables.

But regardless of the method used, you should look for wide differences in the psi readings between cylinders. Low readings mean loss of compression. In some instances you can hear a compression leak or see it at the head—block mating surface. An old test, which unfortunately does not apply to engines with close-mounted radiators and fans, is to move a candle around the head—block joint as the engine is running. Small, almost imperceptible leaks will cause the flame to flicker. Compression leaks into the coolant may show as large bubbles in the header tank.

Coolant leaks can be verified by pressurizing the radiator. Water or antifreeze in the oil can best be determined by laboratory analysis. Large amounts of water show up as gray sludge, which, in extreme cases, is beaten into the consistency of mayonnaise.

## Preparations for Teardown

The external surfaces of the engine must be clean. It is axiomatic that dirt on the outside will find its way to the inside when the castings are parted. The quickest way to clean an engine is with live steam (see Fig. 6-12). Cover all electrical components, and cover or remove pleated-paper air filters. Another method is to use Gunk concentrate cut with kerosene. Gunk is sprayed or brushed on and allowed to stand a few minutes. If you use the former method, work in a well ventilated place and wear a respirator approved for heavy vapor concentrations. Hose the castings down in fresh water and dry with compressed air. In passing, we should note that while compressed air is invaluable in the shop, it must be regulated to 30 psi, both to prevent damage to delicate engine components and to protect the mechanics. Air at more than 30 psi can penetrate the skin and enter the bloodstream.

Carbon tetrachloride should be avoided since the fumes are quite hazardous. Perchlorethylene and trichlorethylene are less toxic and cut grease about as well as carbon tetrachloride. But wear protective clothing and goggles. A peculiar side effect of trichlorethylene (TCE) exposure is

Fig. 6-12. Steam generator. (Courtesy Malsbary Manufacturing Co., Subsidiary of Carlisle Corp.)

"degreaser's flush." After several weeks of contact with TCE, a couple of beers will cause one's face to break out in flaming red blotches. The only cure is to switch solvents—or quit drinking.

Degreasing operations should be segregated from the rest of the shop, and particularly the assembly bays. Benches and tools should be as remote as possible from splatter associated with cleaning. The better shops have an assembly room which is kept almost sterile.

Drain the lube oil and coolant. Disconnect one or both battery cables to disrupt the starting circuit. As a further safety precaution, place the governor control in the no-fuel position. It is suggested that you wear goggles or a face shield. The suggestion is made by insurance inspectors and well meaning doctors of industrial medicine, but is disregarded by most mechanics.

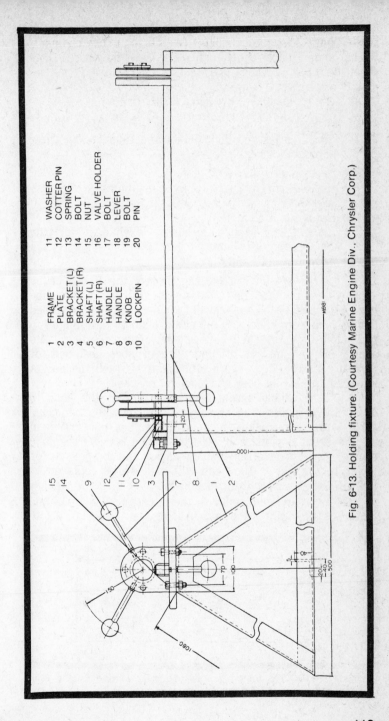

| 1 | FRAME | 11 | WASHER |
| 2 | PLATE | 12 | COTTER PIN |
| 3 | BRACKET (L) | 13 | SPRING |
| 4 | BRACKET (R) | 14 | BOLT |
| 5 | SHAFT (L) | 15 | NUT |
| 6 | SHAFT (R) | 16 | VALVE HOLDER |
| 7 | HANDLE | 17 | BOLT |
| 8 | HANDLE | 18 | LEVER |
| 9 | KNOB | 19 | BOLT |
| 10 | LOCKPIN | 20 | PIN |

Fig. 6-13. Holding fixture. (Courtesy Marine Engine Div., Chrysler Corp.)

113

It is worthwhile to make up a partitioned container for the various parts so that they can be assembled as they come off. Rocker arms, spacers, injectors, push rods, and other fittings should be replaced in their original positions to take advantage of the wear-in which has already occurred. Head work is facilitated by a holding fixture (Fig. 6-13).

**Head Removal**

The drill varies between makes and models, but these parts will have to come off:

- Air cleaner or header.
- Rocker arm cover and (in most instances) rocker arm assembly.
- Exhaust manifold.
- Accessory gear such as the heat exchanger, air trip, and decompressor control lever.
- Attaching bolts and washers.

If you do not have a fixture, protect the fire deck by supporting the head on wooden blocks. The combustion chambers and piston tops should be decoked. Use a dulled knife (a linoleum knife with the end ground at right angles to the handle is ideal) and a power-driven wire brush. Auto supply houses sell stiff-bristled wire brushes designed for carbon removal. Be very careful not to scratch the gasket surfaces. Inspect the head and block for cracks.

Lay a machinist's rule lengthwise across the head to determine the extent of warpage. The allowable warpage varies with engine makes and head dimensions, but 0.005 in. in the longitudinal plane and 0.003 in. in the transverse plane represents more distortion than the gasket can compensate for (refer to Fig. 6-14). The head will have to be replaced or machined if excessive distortion exists.

Milling the head reduces the clearance volume and increases the compression ratio. All engines are designed with

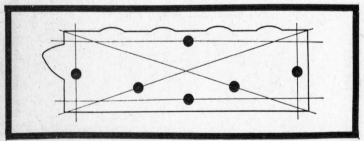

Fig. 6-14. Head measurement points. (Courtesy Marine Engine Div., Chrysler Corp.)

some surplus metal to accommodate light machining, but if the head has already been reconditioned, one may have reached the limit. The specification, of course, varies with manufacturer. As a rough rule of thumb, few heads will tolerate more than 0.020 in. The amount milled should be stamped on the fire deck outside of the sealing surface, or, if that is not possible, on the water jacket.

Injector sleeves generally should be removed before machining the head. One of these sleeves, or *shells*, as they are sometimes called, is shown in Chapter 4 (Fig. 4-6). The sleeve is in direct contact with the coolant. The upper edge is sealed by a neoprene gasket. Special tools are required for removal and installation, and particularly for the final reaming operations. Do not attempt to work injector tubes without the factory aids since inaccuracies in seating can cause coolant leaks and injector failure. An individual or a small shop would be best advised to farm this work out to a facility which specializes in head work.

With injector tubes removed, the ferrous heads can be soaked in a solution of oxalic acid to remove corrosion in the water jacket. Alternatively, the head can be flushed with high-pressure water in the reverse direction of normal circulation flow. Flushing will at least remove the loose corrosion and scale.

The last operation on the head proper is to test it for leaks. Seal off the water holes in the head with made-up plates and rubber gaskets. If injector tubes are present, install dummy injectors to insure proper sealing of the tubes. Tap an air-line connection into one of the expansion plugs and place the head in a container of hot water. The water should be almost boiling. Pressurize the head to between 40 and 100 psi (the specification varies with makes) and allow it to remain in the water bath for a half-hour or so. Assuming that your gaskets are tight and that none of the plugs leaks, bubbles mean that the head is cracked and should be replaced. Heads have been welded and "sewn" with various patented processes. But, at best, these measures are only a stopgap and should not be seriously considered as alternatives to purchasing a new or good used head. Barring a full-fledged leak test, you should at least have the head Magnifluxed.

## Head Assembly

Install new O-rings and *always use a new head gasket*. Sometimes you can get away with using a laminated-steel head gasket twice, but this practice is not recommended. Some mechanics use a sealant such as Kopper-Kote or

Wellseal on the fire deck. Others believe that sealant is the sign of the amateur and only use it on the back surface of oil seals. However, many new engines are assembled with the aid of sealant, so someone likes the stuff. Observe the lay of the head gasket. All bolt holes and ports should be open. Most gaskets are marked FRONT or TOP.

Before the head is in place, take one final look at the gasket, and especially at any O-rings. It is easy to jar one of these rings out of its recess. Lower the head slowly, parallel to the block. Do not skate the head across the gasket.

The capscrews or stud threads should be clean in order to obtain honest torque readings. Burrs and small deformations have a tremendous effect on the torque. Coat the threads with International Compound No. 2, or the equivalent. Tighten the head in several increments in the manufacturer's suggested sequence. A typical sequence is shown in Fig. 6-7. Note the way in which the sequence begins at the center and works outward in a crisscross fashion. This torque sequence illustrates the basic principle of working the fasteners from the center outward so that the head is not buckled between adjacent bolts. But it may be modified in asymmetrical castings or bolt arrangements. Always obtain a copy of the manufacturer's suggested sequence.

Torque limits are a function of bolt material, thread length, and diameter. Exceeding the recommendations will stretch the bolts and give less holding power. The matter of "hot torquing" is subject to dispute. Detroit Diesel recommends that the head fasteners be retorqued after the coolant has reached 160°F. Other manufacturers insist that the engine be allowed to cool down before final torquing.

If the head or block has been machined, the valve clearances will have to be set to an initial value. Failure to make this initial setting can result in valve—piston collision on startup.

## VALVE AND VALVE-GEAR REWORKING

After the rocker arm assembly has been lifted, remove the valves with the aid of a spring compressor. Keep all valves, keepers, springs, and caps in the same sequence so that they can be replaced in the valve from which they were removed.

Valve failure may occur at the seat or along the stem. Less frequently the quill (the recess at the valve stem which locates the keepers) hammers open and allows the valve to drop into the cylinder with dramatic effects. Stem and guide problems may be the result of wear or carbon and gum accumulations which hold the valve open against spring pressure. These deposits are caused by the wrong type of lube for the engine in question, ethylene gylcol leakage into the sump, scoring by foreign matter, and operating at low temperatures (below 160°F). Low temperatures are usually the result of long periods at idle and aggravate any mismatch in the fuel and engine demands. Fuel which burns well at normal temperatures will gum the guides and carbon over the valve heads when the engine is cold. Sticking can also be caused by bent stems. The sealing surface on the valve face is subject to burning and pitting. The last condition is caused by carbon and inadequate lash.

Wire-brush the valve to remove all carbon, and clean the stems with a good commercial solvent. Check for stem trueness by chucking the valve in a lathe or by rolling it on machinist's blocks. Do not attempt to straighten a valve even if you have the prequisite V-blocks and dial indicator. Less than factory trueness may cause the valve to stick and contact the piston.

The inset or extension, as the case may be, of the valves into the combustion chamber is as critical as any engine dimension. Because the lift is fixed by the cam profile and rocker arm proportions, the measurement is taken with the valve seated. Should the valve extend too far into the cylinder it will collide with the piston. Figure 6-15 gives this and other critical dimensions for valves used on Bedford engines. Some designs seat the valve slightly into the combustion space.

Overenthusiastic use of the seat grinder will lower the compression ratio below performance criteria. As far as possible, individual intake and exhaust valves should be seated consistently, to the specified depth for the valve type as supplied by the manufacturer.

High-speed grinders are used to reface the valves and to insure that the face is concentric with the stem. The width of the contact is determined by the seat. In some cases it will be necessary to trim the circumference of the valve to obtain a $^1/_{32}$ in. edge. Narrower edges overheat.

Minor imperfections can be corrected by lapping. However, this method is very limited and time consuming. When significant amounts of metal must be removed, the valve seat is necessarily widened. Wide seats do not seal well, because the area may be perfect at room temperature but leak when the valve becomes hot and "grows."

Lapping compound may be oil- or water-soluble. Most mechanics prefer *Lucky Clover*, or the equivalent, with a petroleum base. The tins are double-walled. One side contains

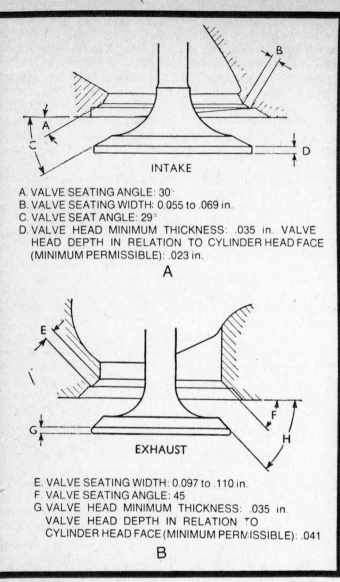

A. VALVE SEATING ANGLE: 30°
B. VALVE SEATING WIDTH: 0.055 to .069 in.
C. VALVE SEAT ANGLE: 29°
D. VALVE HEAD MINIMUM THICKNESS: .035 in. VALVE HEAD DEPTH IN RELATION TO CYLINDER HEAD FACE (MINIMUM PERMISSIBLE): .023 in.

A

E. VALVE SEATING WIDTH: 0.097 to .110 in.
F. VALVE SEATING ANGLE: 45
G. VALVE HEAD MINIMUM THICKNESS: .035 in. VALVE HEAD DEPTH IN RELATION TO CYLINDER HEAD FACE (MINIMUM PERMISSIBLE): .041

B

Fig. 6-15. Valve-machining specifications for current Bedford Engines.

coarse compound for rough cutting, and the other a much finer abrasive for finish work. Some shops dispense with the finish.

A fine coil spring should be placed under the valve to hold it off its seat with a pressure of a few grams. Dab a bit of compound on the face at several places on its circumference. Turn the valve with a suction cup tool held between your palms. The arc should be 100° or so. At intervals move the valve to another spot on the seat, so that the full area is lapped.

Replenish the compound frequently, but do not allow any to dribble down into the guides. Dry lapping will score the seat and face. You should hear a "swish" as the valve is turned.

There are several ways to tell when you are finished. Some mechanics from the old school drop the valve on the seat. A perfectly mated valve gives off a dull thump as it hits. High spots will cause the valve to ring. On recessed heads you can assemble the keepers and spring and fill the recess with kerosene or other light solvent. It should not leak past the valve. The Navy recommends coating the seat with *prussian blue* and giving the valve a quarter-twist. The blue should be broken all the way around the seat.

### Valve Seats

The valve seats or inserts are shrink-fitted into the heads. While this method may not seem reliable, considering the different coefficients of expansion between the heads and the inserts, it does work and has been proven superior to earlier approaches. Seats must be replaced when ground beyond specifications, cracked, or loose. Normally the seat is driven out from below. A few manufacturers supply tools to cut the inserts out (Fig. 6-16). Clean the counterbore with solvent and dry with compressed air. It is always wise to check the counterbore dimensions against specifications, and

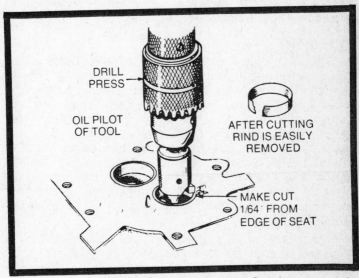

Fig. 6-16. Flycutting seats. (Courtesy Onan.)

particularly so if the seat has been loose. Most suppliers can provide oversized seats. To install, heat the cylinder head in water at between 189° and 200°F for at least a half-hour. This will expand the recess to accept the seat. Place the head face up on the bench and drive the seat home. Work quickly—if you hesitate and allow the temperatures to equalize, the seat will stick or possibly warp. Some mechanics freeze the seats to gain more working time.

Seats are ground with portable tools that center in the valve guide. Detailed instructions accompany these tools: here it is enough to outline the procedure. Begin with the major cut, which is normally 30° or 45°. Open the back of the insert with the 60° wheel and adjust the width of the face with the 15° wheel. These two wheels can be used to raise or lower the insert face relative to the valve face. Ideally, the parts should meet at the center of the valve face.

## Valve Springs

Inspect the springs for evidence of pounding or pitting. If the spring has weakened it is possible that standing waves may flatten the coils upon each other. Spring tension may be measured by a fixture in conjunction with a torque wrench, or measured directly as pounds required to unseat the valve. Manufacturers specify either of these methods, often in conjunction with a measurement of the free length of the spring. If test equipment is not available, it is good practice to replace the springs whenever the valves are disturbed. In some cases weak springs can be shimmed.

## Valve Guides

The wear limit is normally 0.005−0.006 in. Loose guides can make seating difficult and can cause excessive lube-oil consumption. Oil may be drawn into the chamber between the intake valve stems and guides in response to cylinder vacuum during the intake stroke. On 2- and 4-cycles the exhaust side guides may pass oil into the manifold, especially if exhaust back-pressure is low. This condition will be most noticeable at idle. Most designs have a provision for valve guide oil seals in the form of rings below the valve locks. These rings are adequate as long as guide wear is within limits. Some engines may respond to Teflon seals, which require that the guide OD be modified slightly (Fig. 6-17).

Guides are normally driven out from the valve side. It is important to use the correct-sized drift when installing. Otherwise, the guide will bell or band, and it may cause the valve to stick. The depth of the guide is fixed by a counterbore

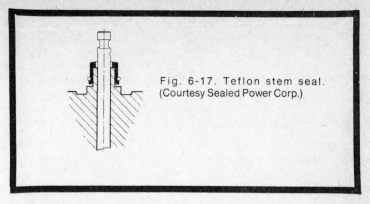

Fig. 6-17. Teflon stem seal.
(Courtesy Sealed Power Corp.)

on most modern engines; on engines without this feature, drive the new guide to the depth of the originals.

### Push Rods

Inspect the push rods for wear on the tips and for bends. The best way to determine their trueness is to roll the rods on a machined surface. Sight between the rod-and plate. If you can see daylight, the rod is bent, and valve lift is reduced a corresponding amount.

### Rocker Arms

The rocker arms are secured by pedestals and are located by silencing springs on the rocker arm shaft. Note the location of any spacers and wave washers. The conical springs may have wide coils at one end which contact the spacers.

Mark the rocker arms as a guide to assembly and inspect the following:

- Adjusting screws: Check the thread fit, and particularly the flat or spherical end which contacts the push rods. Adjusting screws are only surface hardened, and wear past the "skin" will mushroom them, making it impossible to keep the valves adjusted. Replace as needed.

- Rocker arm faces: The faces take on a definite wear pattern as they force the valves open. With the proper equipment, you can regrind the faces to their original contour. The reward is an exceptionally quiet engine.

- Flanks of the rocker arm: Check for cracks radiating out from the fulcrum. This sort of failure is rare but by no means unknown in well used engines. Shops which give extended warranties on their work sometimes Magniflux the rocker arms along with other critical parts.

- Rocker arm bushings: Clearance between rocker arm bushings and an unworn part of the shaft should be on the order of 0.002 in. The old bushings should be pressed or driven out with a suitable punch. Use a vise or accurately sized drift to drive the new one in place. Observe the position of the oil port. On some designs the split in the bushing is part of the oil circuit and aligns with the oil port. Ream to specifications.

The rocker arm shaft wears at the bottom. Compare the journal areas with unworn portions of the shaft. If the shaft is part of the oil supply system, blow out the ports and test for flow with compressed air.

**Cam Followers**

Cam followers, or tappets, are usually not disturbed during cylinder head removal. The followers are buried in the block or, at best, are accessible at the side plates. The GM 2-cycle family is an exception. Its roller tappets are integral with the valve and injector pump gear. Figure 6-18 shows the maximum allowed runout and end clearance for the roller. Inspect for scuffing, flat spots, and ease of rotation. If the roller is damaged, inspect the cam which operates it. Replacement roller and pin sets are available from the factory.

Oversized pins and rollers are also available as installed on cam followers. These are identified by the code letter *S*. Do

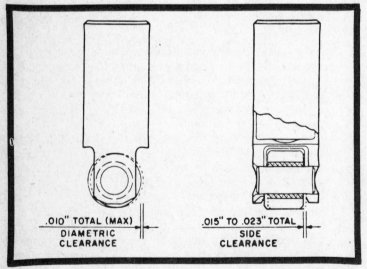

Fig. 6-18. Wear limits of roller tappet. (Courtesy General Motors Corp.)

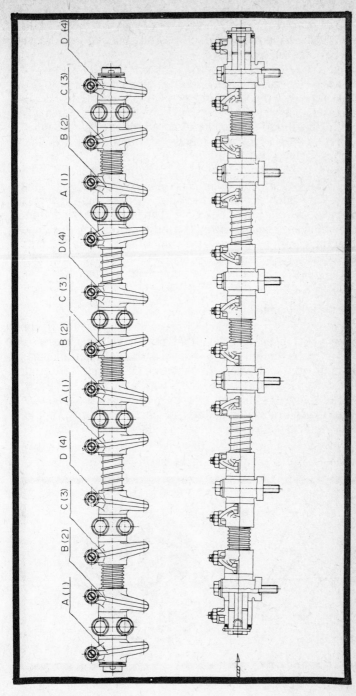

Fig. 6-19. Assembly sequence for rocker arm. (Courtesy Marine Engine Div., Chrysler Corp.)

123

not attempt to ream cam follower legs to accept these larger pins. The job cannot be done successfully outside the factory because of the tolerances involved.

Initial lubrication is extremely important. During the first few seconds of running, the only lubrication the follower will have is what you provide. Lube oil is too viscous to penetrate immediately between the roller and pin. Fuel oil is positively harmful and almost guarantees a scored roller or pin. Remove the preservative from new followers with Cindol 1705, and clean used ones with the same preparation. Just before installation soak the followers in a bath of heated Cindol (100°–125°F). Turn the rollers to release trapped air. Upon removal, air bubbles which are still in the bearing will shrink and draw the Cindol into the cavity. The oil port at the bottom of the follower is pointed away from the valves. There should be 0.005 in. clearance between the cam follower legs and guide. The easiest way to make this adjustment is to loosen the bolts slightly and tap the ends of the guide with a brass drift. The bolts should be torqued to 12–15 ft-lb.

**Valve Gear Assembly**

Assembly is quite straightforward (Fig. 6-19). If the brackets have been disturbed, they may have to be heated to 180°F in water or oil to slip over the shaft. Clean all parts thoroughly and lubricate with generous quantities of lube oil. You can lubricate the valve stems with a mixture of oil and graphite. Before tightening the rocker arm pedestals, be sure the push rods are seated on their respective tappets. Tighten the rocker arm bolts to specifications; do not overtighten.

With valves up, check the end clearance on the two outboard rockers. A ballpark figure is 0.020 in. (Fig. 6-20).

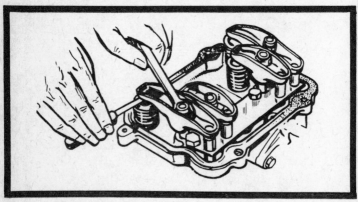

Fig. 6-20. Valve lash adjustment. (Courtesy Onan.)

Make the initial valve lash setting to preclude any possibility of a piston striking a valve which is held off its seat. On 4-cycles this preliminary adjustment and the final hot adjustment can be made more methodical if you bar the engine over until the valves overlap on the last cylinder—that is, so that the exhaust valve is about to close and the intake valve is just beginning to open. The table below shows that the overlap is a narrow image of the firing order. Adjust from the push rod (Detroit Diesel), at the rocker arm, or at the rocker arm stud (Fig. 6-20).

### Four-Cylinder

| Overlap position on | Adjust lash on |
| --- | --- |
| 4 | 1 |
| 2 | 3 |
| 1 | 4 |
| 3 | 2 |

### Six-Cylinder

| Overlap position on | Adjust lash on |
| --- | --- |
| 6 | 1 |
| 2 | 5 |
| 4 | 3 |
| 1 | 6 |
| 5 | 2 |
| 3 | 4 |

# Engine Block Service

**7**

Diesel blocks are, with few exceptions, made of heavy-section cast iron, liberally webbed to withstand large operating stresses. Most are also strengthened by skirts carried below the main-bearing centerline, as in the photo of Fig. 7-1. In this instance the main-bearing journals are further stabilized by *cross bolting*, a procedure that is not unique to Murphy but has become almost a trademark for the firm. Figure 7-2 is interesting because of the full-circle webs and the fact that the casting is aluminum. The thermal conductivity of a light alloy is advantageous for this air-cooled engine.

## PREPARATION

The British military adage that time spent in reconnaisance is not wasted is very true here. Block work is major work, and not to be entered into blindly or without careful weighing of the alternatives.

Depending upon the degree of wear and corrosion, block assemblies are either *overhauled* or *rebuilt*. (An overhaul can be performed with the engine in its mounts, as long as the mechanic has room to crawl under it.) The precise meaning of these terms varies between shops and according to the customer's ability to pay.

At bare minimum an *overhaul* means new rings and main- and rod-bearing inserts. Most of the mechanic's time should be spent inspecting the wear areas—cylinder bores, timing chains, oil seals, cam and balancer shaft bearings, and oil pump. The shaft and cylinders should be miked, and resurfaced as necessary.

A *rebuild* could mean that every frictional surface was brought back into tolerance by parts replacement or

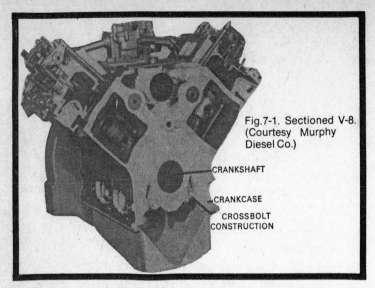

Fig.7-1. Sectioned V-8. (Courtesy Murphy Diesel Co.)

CRANKSHAFT

CRANKCASE

CROSSBOLT CONSTRUCTION

corrective machining. In practice, the average shop rarely does this, but instead follows a schedule of operations. The least that should be done is to bore the cylinders, turn the crankshaft, and replace the cam bearings, oil pump, and all seals. Rebuilt block assemblies are considered high-profit

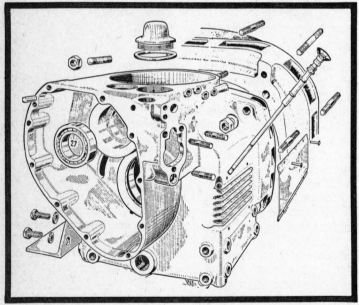

Fig. 7-2. Aluminum block. (Courtesy Teledyne Wisconsin Motor.)

items, but are risky from the shop's point of view since some sort of warranty must be offered.

Many prefer to purchase a block from a rebuilder who specializes in this work and should be able to perform it more reliably than the average shop mechanic. In any event these so-called "factory" rebuilts are at least consistent. Engines from the same outfit will have undergone the same operations. But the skill, tender loving care, and whatever else it takes to make a powerful, reliable engine varies among rebuilders. Some shops do a better job than any "factory" rebuilder. When shopping for a rebuilt or an overhaul, pay more attention to the outfit's reputation than the price.

But no rebuild—not even using a mechanic who learned the trade at Rudolph Diesel's knee—could be the equivalent of a new engine. Theoretically this should be the case; or at least some knowledgeable technicians have argued that it is. In practice rebuilds fall below new engines in terms of reliability (or avoidance of sudden failure) and longevity. Castings suffer fatigue damage; water jackets corrode; and in some instances machining, while acceptable, is a less than perfect solution. For example, boring an engine will increase its compression ratio (CR) since the volume of the cylinder increases as the cube of the bore. A higher CR means easier cold weather starting and more torque. But it also means additional heat transfer to the cooling system and greater bearing loads, along with the possibility of upsetting the balance factor with the necessarily heavier pistons. While this and other modifications may be acceptable, they are not what the designers intended. The better they did their work, the more carefully they made the series of compromises that comprise the art of design; and the more demands they made on the foundry and the machinists, the less room there is to alter their design.

An original short block—one from the same factory which made the engine—is technically the best solution to heavy wear or damage. Short blocks come in several varieties, depending on the manufacturer. The basic package consists of a new block casting, crankshaft, connecting rods, pistons, oil pump, and accessory drives. All the mechanic has to do is to bolt up the sump, head, and accessories. If the head work has been done properly and the pump and other accessories have been dismantled and inspected, the engine will be a close facsimile to a brand new one. But the cost will also approach that of a new replacement.

## DIAGNOSIS

Before you begin you should have a good idea—or at least a plausible theory—about the problem. Consult the engine log.

Comparing operating hours against maintenance operations will not tell you what is wrong, but it will provide background that can be useful.

For example, if the engine develops a knock within 50 hours of the last overhaul, you can be almost certain that the problem is localized in one cylinder, or at least in a single component in the accessory drive. And you can be almost as sure that whoever did the overhaul forgot something like a lockwasher or a torque wrench. On the other hand, knocking at 3000 hours bodes ill since it is probably symptomatic of generalized wear. The log should record fuel and oil consumption. Gradual increases in these figures are normal, but a sudden variation can mean trouble. An engine which stops using oil has not given itself a new set of rings. Something is wrong. The sump could have been overfilled, the dipstick misread because of engine tilt, or, what is more likely, the lube oil has been displaced by fuel oil. Excessive consumption may well be a more serious problem, if it develops slowly over time. A suddenly developed appetite for lube oil usually means a leak or failure of a single component.

Make as many tests as you have instrumentation for. The tests will not only help you localize the problem, but will also serve as a means to measure the effectiveness of your work. Run the engine under a known load and note the rpm drop. Make a compression check of each cylinder as per manufacturer's specifications. Note the quality and color of the exhaust smoke. Blue smoke means that oil is being burned. It could come from a leaking turbocharger seal, an overfilled air filter case, or from oil seals leaking at the valves. The usual route is past worn rings. Consumption is further increased in a worn engine by oil which bleeds out from the rod bearings and is flung onto the cylinder walls.

The surest guide to engine condition is to mail an oil sample—a couple of ounces, taken from the input line to the filter—to a testing lab for spectroscopic analysis. The sample should be taken from a warm engine, immediately upon shutdown.

## SPECTROSCOPIC ANALYSIS

Oil testing was begun by the railroads with their diesel locomotives, but remained somewhat primitive until the U.S. Navy got into the act. These early tests could detect bearing failure, but the Navy—principally at the testing station in Pensacola, Florida, made oil testing a predictive instrument. By the middle 1950s Navy engineers could forecast engine failure at some more or less precise time in the future. They could determine the exact condition of the engine relative to wear and corrosion damage. This program saved the fleet air arm millions of dollars and more than a few pilot's lives. Since then it has been used widely in industry and the service trades.

The trace materials found in lube oil and their sources are as follows:

Lead . . . . . . . . . . . . . . . . . . . . . . . . . . . . . . . . . . . . . . . bearings
Silver . . . . . . . . . . . . . . . . . . . . . . . . . . . . . . bearings
Tin . . . . . . . . . . . . . . . . . . . . . . . . . . . . . . . . . . . . . . . bearings
Aluminum . . . . . . . . . . . . . . . . . . . . bearings, pistons
Copper . . . . . . . . . . . . . . . . . . . . . . . . . . . . . . . . . bushings
Iron . . . . . . . . . . . . . . . . . cylinder bores, piston rings
Chromium . . . . . . . . . . . . . . . . . . . . . . . . . . . . . . . . . rings
Nickel . . . . . . . . . . . . . . . . . . . . . . . bearings, valves

In addition silicon and aluminum traces can be found which indicates a failure of the air filter elements. Other metallic elements—such as boron, zinc, and calcium—are present as oil additives, but in high concentrations can be considered contaminants.

Spectroscopic analysis involves vaporizing the oil sample in an electric arc. Each element has its own characteristic signature in the sense that it gives off light at a single frequency. The light beam is directed through a prism to sort out the frequencies, which are then directed into a bank of photocells. The photocells generate small currents in direct proportion to the intensity of the light. The contaminant level is read directly on meters or from computer printouts. Modern spectrometers can detect traces as fine as one part per million.

A typical oil test requires that 16 substances be identified, and may be repeated several times. The raw data is turned over to an analyist who, on the basis of other information about the engine (its service history, use profile, and the like), makes the final report. If the analyst discovers that the engine is about to blow, he contacts you by phone. The whole thing sounds frightfully expensive, but Analysts, Inc. (820 E. Elizabeth Ave., Linder, N.J. 07036) will do the job for less than $15.

Ideally, you should have the oil tested at least once a year. The reports are much more meaningful when data is cumulative. But a one-shot test can direct your efforts to the trouble spots.

Assuming that the engine will be torn down completely, you will need a chain fall to lift the block, and some sort of engine stand. Take a little time to design a proper hoist

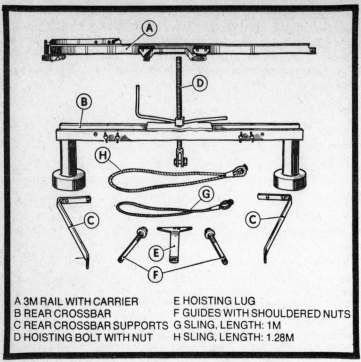

A 3M RAIL WITH CARRIER    E HOISTING LUG
B REAR CROSSBAR    F GUIDES WITH SHOULDERED NUTS
C REAR CROSSBAR SUPPORTS    G SLING, LENGTH: 1M
D HOISTING BOLT WITH NUT    H SLING, LENGTH: 1.28M

Fig. 7-3. Lifting apparatus. (Courtesy Peugeot.)

apparatus since many older engines do not have lifting lugs. Wrapping a couple of turns of chain around the block may be quick, but it can be dangerous to the mechanic and may damage the headers or other parts. Figure 7-3 illustrates a rather complex lift mechanism, used on one Peugeot engine. You probably don't need anything as elaborate as this, but its complexity underlines the seriousness of the problem. An engine stand is almost a necessity. It should be the side mounting type, with floor wheels and a lift mechanism on the pattern of IHC's No. FES-52.

## TEARDOWN

Drain the oil and coolant and degrease the outer surfaces of the engine. Disconnect the battery, wiring harness (make a sketch of the connections if the harness is not keyed for proper assembly) and exhaust system. Attach the sling and undo the drive line connection and the motor mounts. With the engine secured in a stand, detach the manifolds, cylinder heads, and oil sump. The block should be stripped if you contemplate machine work or chemical cleaning of the jackets.

## LUBRICATION SYSTEM

The first order of business should be the lubrication system. To check it out you must have a reasonably good notion of the oiling circuits. Figure 7-4 is a drawing of the Onan DJ series lubrication system. The crankcase breather is included because it has much to do with oil control. Should it clog, the engine will leak at every pore. Oil passes from the screened pickup tube (suspended in the pan) to the pump, which sends it through the filter. From there the filtered oil is distributed to the camshaft, the main bearings (and through rifle-drilled holes in the crankshaft to the rod bearings and wristpin), and to the valve gear. Valve and rocker arm lubrication is through a typically Onan "showerhead" tube. Tiny holes are drilled in the line and deliver an oil spray at about 25 psi. On its return to the sump, the oil dribbles down the push rods to lubricate the cam lobes and tappets.

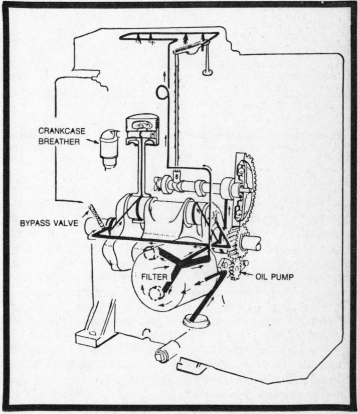

Fig. 7-4. Lubrication system. (Courtesy Onan.)

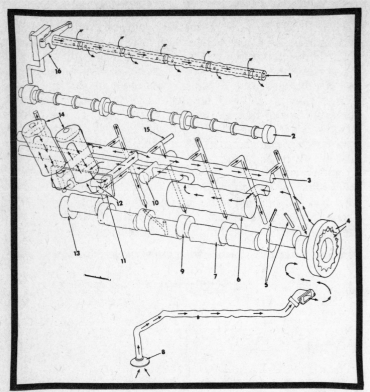

Fig. 7-5. A more complex lubrication system. (Courtesy International Harvester.)

The system in Fig. 7-5 is employed in 6-cylinder engines. From the bottom of the drawing, oil enters the pickup tube then goes to the Gerotor pump. Unlike most oil pumps, this one is mounted on the end of the crankshaft and turns at engine speed. The front engine cover incorporates inlet and pump discharge ports. Oil is sent through a remote cooler (6), then directed back to the block, where it exists again to the filter bank (11). Normally oil passes through these filters. However, if the filters clog or if the oil is thick from cold, a pressure differential type of bypass valve (12) opens and allows unfiltered flow.

The main oil gallery (3) distributes the flow throughout the engine. Some goes to the main bearings and through the drilled crankshaft to the rods. The camshaft bearings receive oil from the same passages which feed the mains. The rearmost cam journal is grooved. Oil passes along this groove and up to the rocker arms through the hollow rocker shaft. On its return this

134

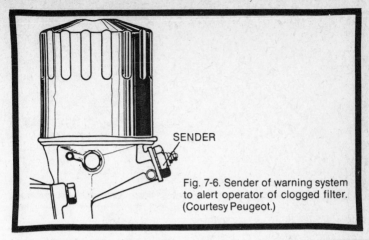

SENDER

Fig. 7-6. Sender of warning system to alert operator of clogged filter. (Courtesy Peugeot.)

oil lubricates the valve stems, push-rod ball sockets, tappets, and cam lobes.

Other makes employ similar circuits, often with a geared pump.

### Filters

Most small diesels have *full-flow filtration*. That is, all the oil which leaves the pump must go through the filter, which is in series with the main gallery. Pleated-paper filters can trap particles as tiny as 1 micron but are handicapped by limited capacity. Should the filter clog, a bypass valve opens, shunting the filter. The valve layout is similar to the one discussed above. Unless the operator takes the trouble to feel the surface of the filter, it is impossible to know if the bypass has opened. Peugeot has solved this problem with a warning light which alerts the driver that the engine is circulating raw, unfiltered oil (Fig. 7-6).

The filter should be changed per the engine maker's instructions, which are more severe than for equivalent SI engines. Because of their higher compression ratios, diesels are more liable to blowby of combustion gases past the piston rings.

Chrysler-Nissan marine plants incorporate a regulating valve at the filter mount, a filter bypass valve, and an oil cooler bypass. The latter cuts the oil cooler out of the circuit until the oil has warmed. Periodically disassemble, clean, and inspect these valves (Fig. 7-7).

### Lube Oil

Oil collects moisture from condensation in the sump and is the respository for the liquid by-products of combustion. These

135

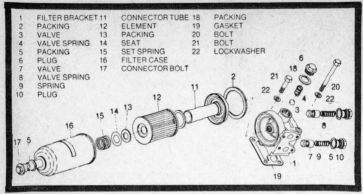

| | | | | | | |
|---|---|---|---|---|---|---|
| 1 | FILTER BRACKET | 11 | CONNECTOR TUBE | 18 | PACKING |
| 2 | PACKING | 12 | ELEMENT | 19 | GASKET |
| 3 | VALVE | 13 | PACKING | 20 | BOLT |
| 4 | VALVE SPRING | 14 | SEAT | 21 | BOLT |
| 5 | PACKING | 15 | SET SPRING | 22 | LOCKWASHER |
| 6 | PLUG | 16 | FILTER CASE | | |
| 7 | VALVE | 17 | CONNECTOR BOLT | | |
| 8 | VALVE SPRING | | | | |
| 9 | SPRING | | | | |
| 10 | PLUG | | | | |

Fig. 7-7. SD22 filter and valve array. (Courtesy Marine Engine Div., Chrysler Corp.)

by-products include several acid families which, even in dilute form, attack bearings and friction surfaces. No commercially practical filter can take out these contaminants. In addition, oil, in a sense, wears out. The petroleum base does not change, but the additives become exhausted and no longer suppress foam, retard rust, and keep particles suspended. Heavy sludge in the filter is a sure indication that the change interval should be shortened, since the detergents in the oil have been exhausted. In heavy concentrations water emulsifies to produce a white, mayonnaise-like gel which has almost no lubrication qualities.

Oil which has overheated oxidizes and turns black. (This change should not be confused with the normal discoloration of detergent oil.) If fresh oil is added, a reaction may be set up which causes the formation of hard granular particles in the sump, known as "coffee grounds."

Oil change intervals are a matter of specification—usually at every 100 hours—and sooner if the oil shows evidence of deterioration. Most manufacturers are quite specific about the type and brand to be used. Multigrade oils—e.g., 10W-30—are not recommended for some engines, since it is believed they do not offer the protection of single-weight types. Other manufacturers specify SE grades. As a practical matter this specification means that multigrade oils can be used. Brand names are important in diesel service. Compliance with the standards jointly developed by American Petroleum Institute, the Society of Automotive Engineers, and the American Society for Testing and Materials is voluntary. You have no guarantee that Brand X is the equivalent of Brand Y, even though the oil may be labeled as meeting the same API-SAE-ASTM standards. GM suggests that you discuss your lubrication needs with your supplier and use an oil which has been successful in diesel engines and which meets the pertinent military standards. MIL-L-210B is the standard most often quoted, although some late oils have been "doped" with additives to meet the API standards. These additives have caused problems with some engines, particularly in deposits around the ring belt. For GM 2-cycles zinc should be held to between 0.07% and 0.10% by weight, and sulfated ash to 1.0%. If the lubricant contains only barium detergent-dispersants, the sulfated-ash content can be increased by 0.05%. High-sulfur fuels may call for *low-ash series 3* oils, which do not necessarily meet military low-temperature performance standards. Such oils may not function as well as MIL-L-2104B lubricants in winter operation.

## Oil Pumps

Oil pumps deserve all the cold-eyed scrutiny a banker would give to an application for a third mortgage. The Gerotor pump shown in Fig. 7-5 is crankshaft driven. Both rotors turn; the inner rotor is mounted slightly off the center of the outer element, and climbs up on it as the pump turns. This movement, somewhat reminiscent of a rotary engine, is possible because the inner rotor has 13 teeth to the outer's 14. Check the radial clearance—the clearance between the outer rotor and the case—with a feeler gage. A good working figure is 0.007 in.

The clearance between the rotors and the end cover is best checked with *Plasti-Gage*. Plasti-Gage consists of rectangular-section plastic wire in various thicknesses. A length of the wire is inserted between the rotors and cover. The working clearance is a function of how much the wire is compressed under assembly torque. The package has a scale printed on it to convert this width to thousandths of an inch. Plasti-Gage is extremely accurate and is fast. The only precautions in its use are that the parts must be dry, with no oil or solvent adhering to them, and that the wire must be removed after measurement. Otherwise it may break free and circulate with the oil, where it can lodge in a port. Typical end clearance for a Gerotor pump is on the order of 0.003 in. Of course, all rubbing surfaces should be inspected for deep scratches, and the screen should be soaked in solvent to open the pores.

Camshaft-driven gear-type pumps are the norm (Fig. 7-8). Check the pump for obvious damage—scores, chipped teeth, noisy operation. Then measure the clearance; the closer the better as long as the parts are not in physical contact.

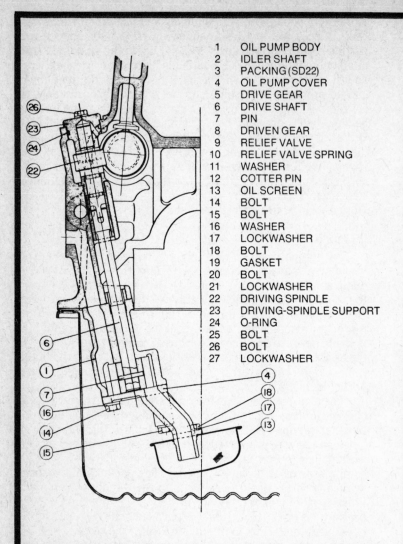

| | |
|---|---|
| 1 | OIL PUMP BODY |
| 2 | IDLER SHAFT |
| 3 | PACKING (SD22) |
| 4 | OIL PUMP COVER |
| 5 | DRIVE GEAR |
| 6 | DRIVE SHAFT |
| 7 | PIN |
| 8 | DRIVEN GEAR |
| 9 | RELIEF VALVE |
| 10 | RELIEF VALVE SPRING |
| 11 | WASHER |
| 12 | COTTER PIN |
| 13 | OIL SCREEN |
| 14 | BOLT |
| 15 | BOLT |
| 16 | WASHER |
| 17 | LOCKWASHER |
| 18 | BOLT |
| 19 | GASKET |
| 20 | BOLT |
| 21 | LOCKWASHER |
| 22 | DRIVING SPINDLE |
| 23 | DRIVING-SPINDLE SUPPORT |
| 24 | O-RING |
| 25 | BOLT |
| 26 | BOLT |
| 27 | LOCKWASHER |

Fig. 7-8. Exploded view of typical gear-type oil pump. (Courtesy Marine Engine Div., Chrysler Corp.)

End clearance can be checked by bolting the cover plate up with Plasti-Gage between the cover and gears, or with the aid of a machinist's straightedge. Lay the straightedge on the gear case and measure the clearance between the top of the gear and the case with a feeler gage. Maximum allowable clearance is 0.005—0.007 in. When in doubt, replace or

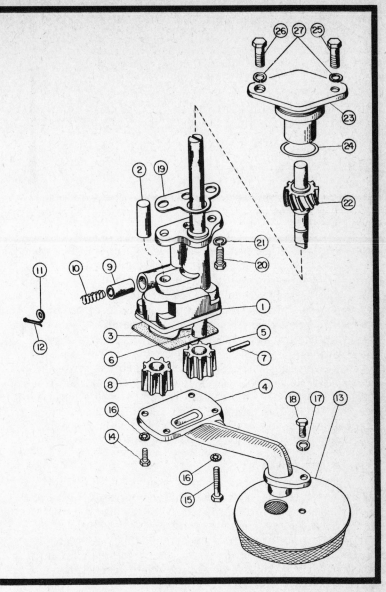

resurface the cover. Check the diameter of all shafts and replace as needed. The idler gear shaft is typically pressed into the pump body. Use an arbor press to install, and make certain that the shaft is precisely centered in its boss. Allowing the shaft to cant will cause interference and early failure. Total wear between the drive shaft and bushing can be

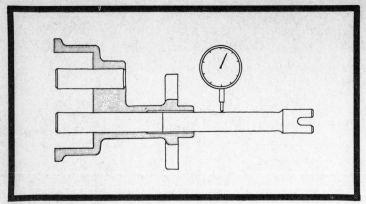

Fig. 7-9. Determining shaft—bearing wear. (Courtesy Marine Engine Div., Chrysler Corp.)

determined with a dial indicator as shown in Fig. 7-9. Move the shaft up and down as you turn it. Check the backlash with a piece of solder or Plasti-Gage between the gears. For most pumps this clearance should not exceed 0.018 in.

In general, it is wise not to tamper with the pressure relief valve, which may be integral with the pump or at some distance from it. If you must open the valve, either because of excessive lube oil pressure or low pressure, observe that the parts are generally under spring tension and can "explode" with considerable force. Check the spring tension and free length against specs, and replace or lap the valve parts as needed. When this assembly has been disturbed it is imperative that the oil pressure be checked with an accurate instrument such as the gage shown in Fig. 7-10.

## Oil Distribution System

Oil galleries, lines, and ports should be flushed with solvent or fuel and blown out with compressed air. Pay particular attention to oil spray jets. Oil spray may be directed at the accessory gear train at the front of the block, at the valves and rocker arms from an overhead tube, or directed at the backside of the piston crowns. Clean the jets with air and a length of straw. If you use wire or a torch tip cleaner, be very careful not to increase the size of the orifices.

As each lubricated part is disassembled, check the associated oil circuit. You can sometimes trace the circuit and discover stoppages and leaks with the help of high-pressure air. Do not, however, play the air jet over your skin. It has been determined that air at 30 psi and over can penetrate human skin and form bubbles in the circulatory system.

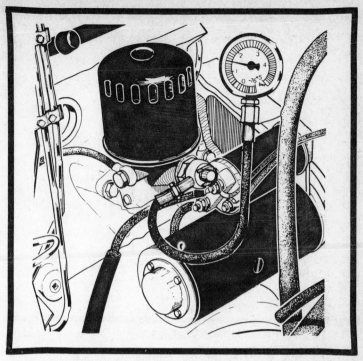

Fig. 7-10. Oil pressure check. (Courtesy Peugeot.)

## Oil Pressure Monitoring Devices

Since oil pressure is so critical, it is wise to invest in some sort of alarm to supplement the usual gage. Gage failure is rare (erratic operation of mechanical gages is usually caused by a slug of hardened oil impacted at the gage fitting), but it does happen. Failure to shut down when pressure is lost will destroy the engine in seconds.

**Steward-Warner Monitor.** Stewart-Warner makes an audio-visual monitor which lights and buzzes when oil pressure and coolant temperatures exceed safe norms. The indicator is a cold-cathode electron tube, which is much more reliable than the usual incandescent bulb. Designed for panel mounting, the 366-TJ retails for less than $20. Onan engines, among others, can be fitted with automatic shutdown circuits should oil pressure drop. The original-equipment kit consists of a pressure sensor, which opens at 13 psi and deenergizes the fuel solenoid, stopping the engine. The starter switch shunts the sensor. Add-on kits are somewhat more elaborate.

**Onan Monitor.** Onan fits a *normally on* sensor element and supplies a time delay relay. In conjunction with a 10W, 1-ohm

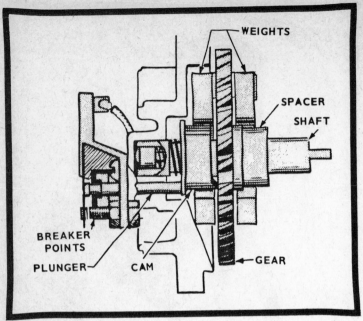

Fig. 7-11. Sensor of Onan oil pressure monitor.

resistor. this relay allows the engine to be started. When running pressure drops below 13 psi, the sensor opens, causing the fuel rack to pull out.

The sensor (Fig. 7-11) is the most complicated part of the system. It has a set of points which should be inspected periodically, cleaned, and gapped to 0.040 in. When necessary, disassemble the unit to check for wear in the spacer, fiber plunger, and spring-loaded shaft plunger. The spacer (shown in Fig. 7-12) must be at least 0.35 in. long. Replace as needed. Check the action of the centrifugal mechanism by moving the weights in their orbits. Binding or other evidence of wear dictates that the weights and cam assembly be replaced. The cam must not be loose on the gear shaft.

**Ulanet Monitor.** The George Ulanet Company makes one of the most complete monitoring systems on the market (Fig. 4-12). It incorporates an optional water level sensor, overheat sensor, fuel-pressure sensor, and low- and high-speed oil sensors. For the GM 3-71 and 4-71 series, these sensors are calibrated at 4 psi and 30 psi, respectively.

### Crankcase Ventilation

The crankcase must be vented to reduce the concentration of acids and water in the oil. A few cases are at atmospheric

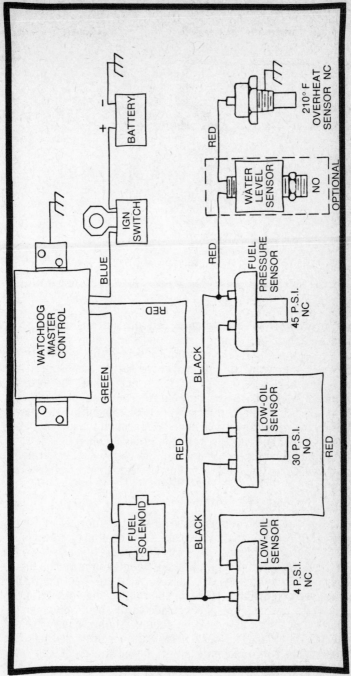

Fig. 7-12. GM 3-71 and 4-71 Watchdog wiring diagram. (Courtesy George Ulanet Co.)

143

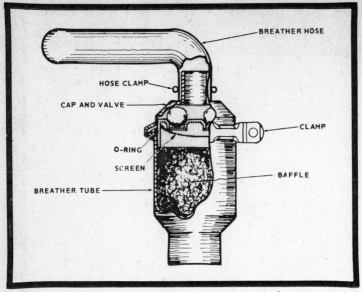

Fig. 7-13. Crankcase vent and valve. (Courtesy Onan.)

pressure; others (employing forced-air scavenging) are at slightly higher than atmospheric; and still other systems run the crankcase at a slight vacuum to reduce the possibility of air leaks. The vapors may be vented to the atmosphere or recycled through the intake ports. In any event, the system requires attention to insure that it operates properly. A clogged mesh element or pipe will cause unhealthy increases in crankcase pressures, forcing oil out around the gaskets and possibly past the seals, as well as increasing oil consumption. Figure 7-13 shows a breather assembly with ball check valves to insure than the case remains at less than atmospheric pressure.

## CYLINDER BLOCK

After the block is degreased, inspect it thoroughly, with particular attention to the lubrication and cooling circuits.

On flat-deck engines check for block distortion as shown in Fig. 7-14, using a machinist's straightedge and a feeler gage. Generally, new engines are held to 0.002 in. The head and gasket are conformable to distortion in the range of 0.06 in. If the block is severely warped, consideration should be given to replacing it, although most manufacturers allow conservative milling. The change in deck height dimension increases the compression ratio and may upset the geometry of vee and slant-block engines.

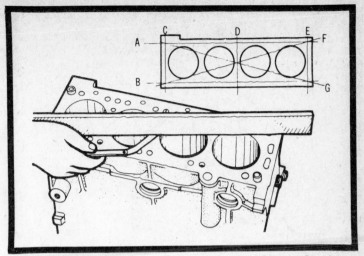

Fig. 7-14. Checking for block distortion. (Courtesy Chrysler Corp.)

## General Reboring

Ascertain the amount of bore wear. Most wear occurs near the top of the cylinder at the extreme of ring travel. This wear is caused by oil starvation in this area and is accelerated by acid attack. A cylinder gage is the most accurate way to determine the wear, taper, and eccentricity. Some idea of bore condition can be had by inserting a ring into the cylinder with the flat of a piston. The difference in ring end gap between the upper and lower portions of the cylinder, as determined by a feeler gage, is an indication of cylinder "growth." However, this method is hardly accurate and should not be substituted for precision measuring tools (Fig. 7-15).

Study the surface of the bore under a strong light. Deep vertical scratches usually indicate that the air filter has at one time failed. The causes of more serious damage—erosion from contact with fuel spray, galling from lack of lubrication, rips and tears from ring, ring land, or wristpin lock failure—will be painfully obvious.

Cylinder bores are divided into two categories. *Dry sleeve* engines may be as cast or have a sleeve (cylinder liner) inserted in the bore, serving as the friction surface for the piston and rings. *Wet sleeve* engines have the sleeve in direct contact with the coolant. In general, the latter type is not bored to correct damage, but replaced along with factory-matched piston set. Dry sleeve engines may be bored and fitted with oversized pistons and rings to the bore limit. When this limit is reached a liner is installed and bored to match piston size.

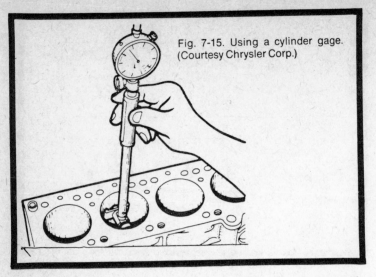

Fig. 7-15. Using a cylinder gage. (Courtesy Chrysler Corp.)

Wet-sleeve blocks simplify foundry work and should result in better control of water jacket dimensions. The sleeves themselves are quite accurately machined. Dry bores are, on balance, somewhat less than perfect in terms of heat transfer, but have several advantages. The bore serves to strengthen the block, and there is no opportunity for coolant leaks into the chamber or sump. On the other hand, heat transfer suffers when a dry liner is used since the interface between the liner and block acts as a heat barrier.

Boring requires precision equipment and the skill to use the equipment. The tool should enter at right angles to the main-bearing centerline and should leave a smooth finish. Obvious threading or "chatter" marks are unacceptable, although reworked cylinders must, in any event, be honed.

How much the bores are opened depends upon the extent of damage and on parts availability. It is always wise to have the machinist be responsible for parts procurement and fitting, since it is the mating of these two which determines how well the job is done. In this country pistons are offered in increments of 0.010 in., with a few manufacturers making 0.005 in. oversizes for slightly cleaned up bores (not for worn bores which have not been honed), and others supplying 0.015 in. piston and ring sets. The wear limit is usually in the range of 0.040—0.060 in., depending upon the cylinder wall thickness and spacing. After the wear limit is reached, the cylinder must be sleeved.

Liners in dry-sleeve engines require a tool for removal and installation, and even slip-fit wet liners tend to stick. The dimensions of the tool vary among engine manufacturers and types. Figure 7-16 shows the general configuration. Before installing, clean the parts with (preferably) one of the solvents used in dry cleaning. Smear the mating surfaces with engine oil.

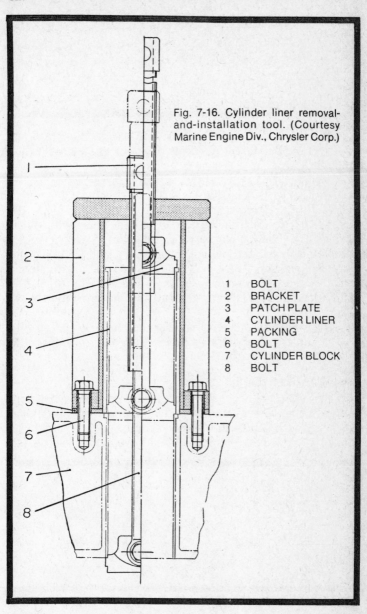

Fig. 7-16. Cylinder liner removal-and-installation tool. (Courtesy Marine Engine Div., Chrysler Corp.)

| | |
|---|---|
| 1 | BOLT |
| 2 | BRACKET |
| 3 | PATCH PLATE |
| **4** | CYLINDER LINER |
| 5 | PACKING |
| 6 | BOLT |
| 7 | CYLINDER BLOCK |
| 8 | BOLT |

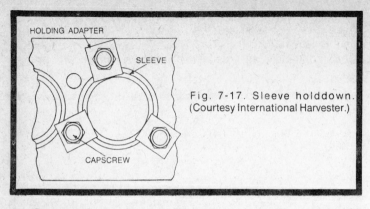

In general, wet sleeves are easier to work, although care is required not to damage the seals. Remove the liners (if they are to be reused, mark them by cylinder number), inspect the counterbore for fractures or burrs, and flush the water jacket. A complication is that the height of the replacement liners must be at a factory-specified dimension above or below the fire deck. Three clamps are used to hold the liner (without the O-ring) into the counterbore. These clamps are generally fabricated in the shop in cold-rolled steel and secured with hardened washers at the bolt heads. The bolts are tightened equally (Fig. 7-17), and the protrusion or depression measured with a dial indicator. Shims are available for installation below the liner lip (Fig. 7-18).

The O-rings and other pliable seals should be coated with petroleum jelly. Install the liners by hand, being careful not to pinch the O-rings or force them out of their grooves.

### Detroit Diesel Reboring

These 2-cycle engines require some special instruction. Series 71 engines may be cast in aluminum. Early production inserts were a slip fit in the counterbore; current standards call for the liner to be pressed in. In any event, the block should

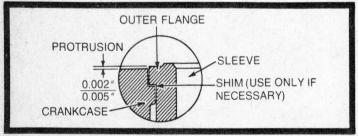

Fig. 7-18. Measuring sleeve protrusion. (Courtesy International Harvester.)

be heated to between 160° and 180°F in a water tank. Immerse the block for at least 20 minutes.

Counterbore misalignment can affect any engine, although mechanics generally believe that the aluminum block is more susceptible to it. You will be able to detect misalignment by the presence of bright areas on the outside circumference of the old liner. The marks will be in pairs—one on the upper half of the liner and the other diagonally across from it, on the lower half. The counterbore should be miked to check for taper and out-of-roundness.

Small imperfections—but not misalignment between the upper and lower deck—can be cleaned up with a hone. Otherwise, the counterbore will have to be machined. Torque the main-bearing caps. Oversize liners are available from General Motors and from outside suppliers such as Sealed Power Corp.

Liners on early engines projected 0.002−0.006 in. above the block (Fig. 7-19). These low-block engines used a conventional head gasket and shims under the liner to obtain flange projection. Late-model *high*-block engines use an insert below the liner. Inserts are available in 0.004 0.008 in. understock thicknesses to compensate for metal removed from the fire deck, and in various oversized diameters to accommodate larger liners. A 0.002 in. shim is also available for installation under the insert.

*Note*: inserts do not last forever. They may become damaged in service and can contribute to upper liner breakage.

Series 53 engines employ a wet liner. The upper portion of the liner is surrounded by coolant and sealed with red silicone seals in grooves on the block. Early engines had seals at the

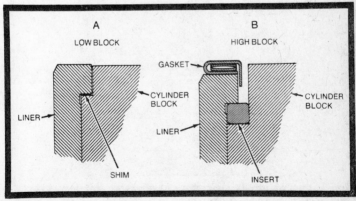

Fig. 7-19. Detroit Diesel liner arrangements. (Courtesy Sepled Power Corp.)

top and above the ports. Late-model engines dispense with the lower seal. A second groove is machined at the top of the cylinder to be used in the event of damage to the original.

The seals must be lubricated to allow the liners to pass over them. Do not presoak the seals, since silicone expands when saturated with most lubricants. The swelling tendency is pronounced if petroleum products are used. Lubricate just prior to assembly with silicone spray, animal fat, green soap, or hydraulic brake fluid. Carefully lower the liners into the counterbores, without twisting the seals or displacing them from their grooves.

The eccentricity (out-of-roundness and taper) must be measured before final assembly. On the 110 series you are allowed a minuscule 0.0015 in. eccentricity. The 53 and 71 engines will tolerate 0.0002 in. Eccentricity can often be corrected by removing the liners and rotating them 90° in the counterbores. Do not move the inserts in this operation.

## Honing

Honing is used to remove small imperfections and glaze. Glaze is the hard surface layer of compacted iron crystals formed by the rubbing action of the rings. Most engine manufacturers recommend that the glaze be broken to aid in ring seating and remove the ridge which forms at the upper limit of ring travel.The Perfect Circle people suggest that honing can be skipped if the cylinder is in good shape.

The pattern should be diamond-shaped, as shown in Fig. 7-20, with the intersection at between 22° and 32° at the horizontal centerline. The cut should be uniform in both

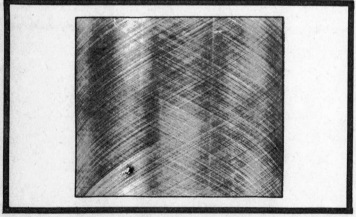

Fig. 7-20. Cylinder honing and crosshatch pattern obtained. (Courtesy Kohler of Kohler.)

directions, without torn or folded metal, leaving a surface free of burnish and imbedded stone particles. These requirements are relatively easy to meet if you have access to an automatic honing machine. However, satisfactory work can be done with a fixed-adjustment hone turned by a drill press or portable drill motor.

The hone must be parallel to the bore axis. Liners can be held in scrap cylinder blocks or in wood jigs. The spindle speed must be kept low—a requirement which makes it impossible to use a quarter-inch utility drill motor. Suggested speeds are shown below.

| Bore Diameter (inches) | Spindle Speed (rpm) | Reciprocations Per Minute |
|---|---|---|
| 2 | 380 | 140 |
| 3 | 260 | 83 |
| 4 | 190 | 70 |
| 5 | 155 | 56 |

Move the hone up and down the bore in smooth oscillations. Do not let the tool pause at the end of the stroke, but reverse it rapidly. Excessive pressure will load the stone with fragments, dulling it and scratching the bore. Flood the stones with an approved lubricant (such as mineral oil) which meets specification 45 SUV at 100° F.

Stone choice is in part determined by the ring material. Most engines respond best to 220—280 grit silicon carbide with code J or K hardness. Oil consumption can be reduced and bore life extended by making 8—15 dwell passes with the stones in very light contact.

Cleaning the bore is a chore which is seldom done correctly. Never use a solvent on a honed bore. The solvent will float the silicon carbide particles into the iron, where they will remain. Instead, use hot water and detergent. Scrub the bore until the suds remain white. Then rinse and wipe dry with paper towels. The bore may be considered "sanitary" when there is no discoloration of the towel. Oil immediately.

**Piston Rings**

Piston rings are primarily seals to prevent compression, combustion, and exhaust gases from entering the crankcase. The principle employed is a kind of mechanical jujitsu—pressure above the ring is conducted behind it to spread the ring open against the cylinder wall. The greater the pressure above the ring, the more tightly the ring is wedged against the wall (Fig. 7-21).

151

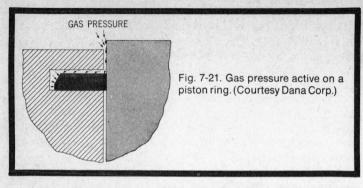

GAS PRESSURE

Fig. 7-21. Gas pressure active on a piston ring. (Courtesy Dana Corp.)

The rings also lubricate the cylinder walls. The oil control ring distributes a film of oil over the walls, providing piston and ring lubrication. One or more scraper rings control the thickness of the film, reducing chamber deposits and oil consumption. In addition to sealing and lubrication, the ring belt is the main heat path from the piston to the relatively cool cylinder.

Rings are almost always cast iron, although steel rings are used in some extreme-pressure situations. Cast iron is one of the very few metals which tolerates rubbing contact with the same material. Until a few years ago rings were finished as cast. Today almost all compression rings are flashed with a light (0.004 in.) coating of chrome. Besides being extremely hard and thus giving good wear resistance, chrome develops a pattern of microscopic cracks in service. These cracks, typically accounting for 2% of the ring's surface, serve as oil reservoirs and help to prevent scuffing. A newer development is to *fill*, or *channel*, the upper compression ring with molybdenum. The outer diameter of the ring is grooved and the moly sprayed on with a hot-plasma or other bonding process. Besides having a very low coefficient of friction and a very high melting temperature, moly gives a piston ring surface which is 15—30% void. It retains more oil than chrome-faced rings and is, at least in theory, more resistant to scuffing.

Rings traditionally have been divided into three types, according to function. Counting from the top of the piston, the first and second ring are *compression* rings, whose task is to control blowby. The middle ring is the *scraper*, which keeps excess oil from the combustion space; and the last ring is the *oil* ring, which is serrated to deliver oil to the bore.

This rather neat classification has become increasingly ambiguous with development of multipurpose ring profiles and the consequent reduction in the number of rings fitted to a piston. Five- and six-ring pistons have given way to three- and four-ring pistons on many of the smaller engines. The function of the middle rings is split between gas sealing and oil control; and the lower rings, while primarily operating as cylinder oilers, have some gas-sealing responsibilities. Design has become quite subtle, and it is difficult for the uninitiated to distinguish between *compression* and *scraper* rings.

The drawing in Fig. 7-22 illustrates the ring profiles used on the current series of GM Bedford engines. Note the differences in profile among the three. These profiles are by no means universal. The Sealed Power Corporation offers several hundred in stock and will produce others on special order.

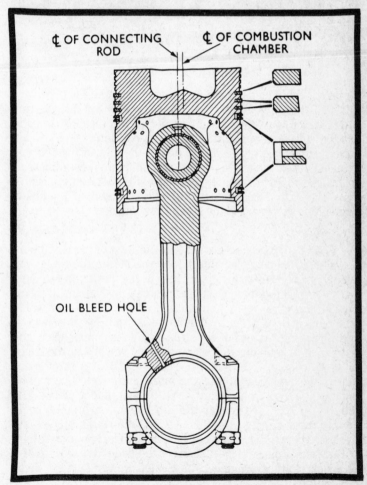

Fig. 7-22. Ring configuration. (Courtesy GM Bedford Diesel.)

Fig. 7-23. Scuffing. (Courtesy Sepled Power Corp.)

What this means to the mechanic is that he must be very careful when installing rings. Most have a definite *up* and *down*, which may or may not be indicated on the ring. Usually the top side is stamped with some special letter code. Great care must be exercised not to install the rings in the wrong sequence. New rings are packaged in individual containers or in groups which are clearly marked *1* (for first compression), *2*, and so forth. Reusable rings should be taken off the piston and placed on a board in the assembly sequence.

### Ring Wear

The first sign of ring wear is excessive oil consumption, signaled by blue smoke. But before you blame the rings, you should check the bearing clearances at the main and crankpin journals. Bearings worn to twice normal clearance will throw off five times the normal quantity of oil on the cylinder walls. You can make a direct evaluation of oil spill by pressurizing the lubrication system. If oil is getting by the rings, the carbon pattern on the piston will be chipped and washed at the edges of the crown.

Check the rings for sticking in their grooves (this can be done on 2-cycles from the air box), breaks, and scuffing. The latter (Fig. 7-23) is by far the most common malady, and results from tiny fusion welds between the ring material and cylinder walls. Basically it can be traced to lack of lubrication, but the exact cause may take all of the deductive talents of a Sherlock Holmes. Engineers at Sealed Power suggest these possibilities:

| Symptom | Possible Cause |
|---------|----------------|
| Overheating | Clogged, restricted, sealed cooling system |
|  | Defective thermostat or shutters |
|  | Loss of coolant |
|  | Detonation |
| Lubrication failure | Worn main bearings |
|  | Oil pump failure |
|  | Engine lugging under load |
|  | Extensive idle |
|  | Fuel wash on upper-cylinder bores |
|  | Water in oil |
|  | Low oil level |
|  | Failure to pressurize oil system after rebuild |
| Wrong cylinder finish | Low crosshatch finish |
|  | Failure to hone after reboring |
| Insufficient clearance | Inadequate bearing clearance at either end of the rod |
|  | Improper ring size |
|  | Cylinder sleeve distortion |

Usually inadequate bearing clearance, complicated by a poor fit in the block counter bore, results in overheating. The fundamental cause is often poor torque procedures, or improperly installed sealing rings on wet-sleeved engines. A rolled or twisted sealing ring can distort the sleeve.

Ring breakage is due to abnormal loading or localized stresses. It can be traced to:

- Ring sticking: This overstresses the free end of the ring.
- Detonation: This is traceable to the overly liberal use of starting fluid, to dribbling injectors, and to out-of-time delivery.
- Overstressing the ring on installation: Usually the ring breaks directly across from the gap.
- Excessively worn grooves: These allow the ring to flex and flutter.
- Ring hitting the ridge at the top of the bore: The mechanic is at fault since this ridge should have been removed.

The last point—involving blame—can be sticky in a shop situation. Mechanics make mistakes the same as everyone else, and the number of mistakes is, in part, a function of the complexity of the repair. Few men can overhaul a machine as

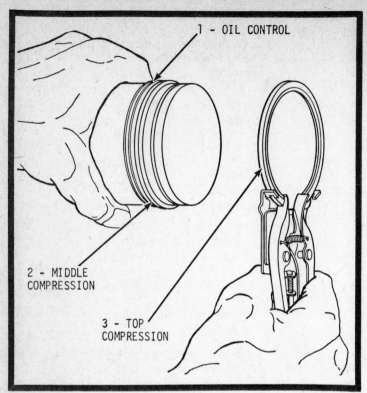

1 - OIL CONTROL

2 - MIDDLE COMPRESSION

3 - TOP COMPRESSION

Fig. 7-24. Piston ring expander. (Courtesy Kohler of Kohler.)

complicated as a multicylinder engine without making some small error. Assessing blame, if only to correct the situation, is sometimes complicated by having the mechanic who built the engine tear it down. But careful examination of the parts usually points to the fault. For example, rings which have been fitted upside down show wear patterns which are the reverse of normal. Rings which have broken in service are worn on either side of the break, from contact with the cylinder walls. The fracture will be dulled. Rings that have been broken during removal show sharp crystalline breaks without local wear spots.

Rings should be handled with care. They are razor sharp and fracture if opened too widely or if twisted. The safest course is to use a ring expander during removal and installation (Fig. 7-24). Open the ring just enough to clear the lands.

Assuming that the rings are the correct diameter and width for the engine, the mechanic only has to make one measurement as far as the rings themselves are concerned. The end gap is critical since too small a gap will allow the end to butt, and too large a gap will increase leakdown. Manufacturers have not standardized the measurement procedure. Some supply specifications on the assumption that the gap will be measured at above ring travel, on the ridge. Others assume that the measurement will be taken at the lower extreme. Consult your shop manual.

Use a piston to guide the ring into place at 90° to the bore axis. Determine the gap with a feeler gage (Fig. 7-25).

The rule of thumb of 0.005 in. per inch of bore diameter has been made obsolete by new ring materials. Ring gaps for the engines in the same general bore classification can vary from 0.015 to 0.025 in. Alloy rings generally have a greater coefficient of expansion than those cast in pure iron.

The width of the piston groove is critical and should be measured with a gage for tapered or keystone grooves (Fig. 7-26), or with a feeler gage for the more conventional parallel grooves. Use a new ring and measure the distance between it and the upper surface of the groove.

Since the rings have residual tension, they must be compressed before installation. GM makes a tapered fitting for their 2-cycles which mates with the sleeve and prevents damage to the rings from excessive compression. The plier-type compressor shown in Fig. 7-27 features interchangeable bands for different bore sizes. Ring gaps

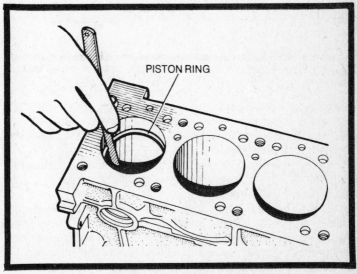

Fig. 7-25. Determining end gap. (Courtesy Chrysler Corp.)

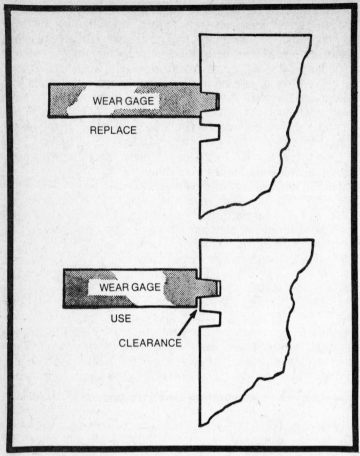

Fig. 7-26. Determining groove wear with a gage. (Courtesy International Harvester.)

should be staggered to prevent blowby during initial starting. Use plenty of lube oil—the surest method is to immerse the pistons past the wristpins in a container of oil and insert the assembly into the cylinder.

Detroit Diesels are "built" on the bench; the piston rod and ring assembly is inserted into the liners, which are then placed into the counterbores.

Drive the piston out of the compressor with a hammer handle (Fig. 7-28). Do not force. The piston should move with each blow. If it stops, one of the rings escaped the compressor and is hanging on the edge of the bore. Should this happen, pull the piston, get a fresh bite with the compressor, and repeat the process.

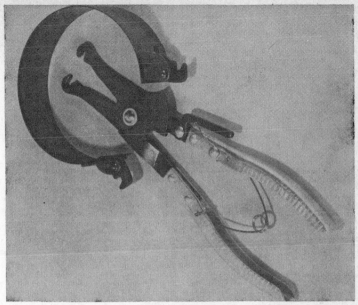

Fig. 7-27. Ring compressor. (Courtesy K-D Manufacturing Co.)

Once installed, torque up the rods and bar the crank over to be certain that none of the rings are sticking.

## PISTONS

Diesel pistons are made of aluminum or cast iron. Die castings have reduced the cost of aluminum pistons; however,

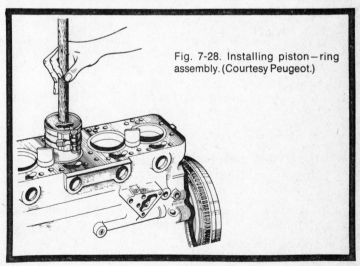

Fig. 7-28. Installing piston—ring assembly. (Courtesy Peugeot.)

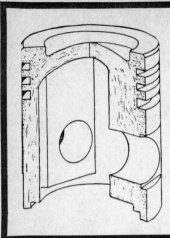

Fig. 7-29. Forged piston showing grain. (Courtesy Sealed Power Corp.)

forgings are preferred for serious work. Forgings are tougher because of the compacted grain structure (Fig. 7-29) and impose little or no weight penalty.

Piston failure is usually quite obvious. Wear should not be a serious consideration in low-hour engines, since the skirt areas are subject to relatively small forces and have the benefit of surplus lubrication. Excessive wear can be traced to dirty or improperly blended lube oil or inadequate air filtration. Poor cylinder finishing may also contribute to it. Piston collapse or shrinkage is usually due to overheating. If the problem shows itself in one or two cylinders, expect water jacket stoppage or loose liners.

*Detonation* damage begins by eroding the crown, usually near the edge. The erosion spreads and grows deeper until the piston "holes." Typically a piston holed from detonation will look as if it were struck by a high-velocity projectile. *Scuffing* and *scoring* (a scuff is a light score) may be confined to the thrust side of the piston. If this is the case look for the following:

- Oil pump problems—screen clogged, excessive internal clearances.
- Insufficient rod bearing clearances, which reduce throwoff, robbing the cylinders of oil.
- Lugging.

The probable causes of damage to both sides of the skirt include the ones just mentioned, plus these:

- Low or dirty oil.
- Detonation.

160

- Overheating caused by cooling system failure.
- Coolant leakage into the cylinder.
- Inadequate piston clearance.

Scuffs or scores fanning out 45° on either side of the pinhole mean one of these conditions:

- Pin fit problems—too tight in the small end of the rod or in the piston bosses.
- Pinhole damage (see below for installation procedures).

Ring land breakage can be caused by the following:

- Excessive use of starting fluid.
- Detonation.
- Improper ring installation during overhaul.
- Excessive side clearance between the ring and groove.
- Water in cylinder.

Free-floating pins sometimes float right past their lockrings and contact the cylinder walls. Several causes (listed below) have been isolated.

- Improper installation: Some mechanics force the lockrings beyond the elastic limit of the material. In a number of cases, it is possible to install lockrings by finger pressure alone.
- Improper piston alignment: This may be caused by a bent rod or inaccuracies at the crankshaft journal. Throws which are tapered or out of parallel with the main journals will give the piston a rocking motion which can dislodge the lockring. Pounding becomes more serious if the small-end bushing is tight.
- Excessive crankshaft end play: Fore-and-aft play is transmitted to the lockrings and can pound the grooves open. Again, a too-tight fit at the connecting rod's small end will hasten piston failure.

**Piston Servicing**

Remove the piston from the rod. Free-floating pins should be removed in either of two ways, depending upon the tools available and mechanic's preference. Perhaps the safest method is to heat the piston in oil and, with the lockrings removed, push the pin through with a wood or soft brass drift. The technique involves a factory-supplied drift and a soft-faced V-block.

The piston should be cleaned in a chemical which will not damage it or its tin plating, if present. However, these cleaners are slow, and most mechanics find themselves using

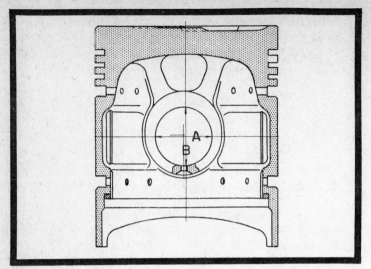

Fig. 7-30. Measuring points of piston pinhole. (Courtesy Marine Engine Div., Chrysler Corp.)

a groove scraper to speed the process. You can purchase one of these tools, or make one up from a broken ring and file handle. The cutting face of the ring should be beveled slightly on a wheel to prevent damage to the grooves. All carbon must be removed and the oil ports opened to their original diameter.

Inspect the piston carefully, checking the groove widths (the upper compression ring groove takes the worst beating), damage to the crown, and cracks radiating from the pin bosses. Scrutinize the pin journal lockring grooves. Check the journal diameter with an inside mike or plug gage and check the piston diameter against specifications. Typical measurement locations are given in Figs. 7-30 and 7-31.

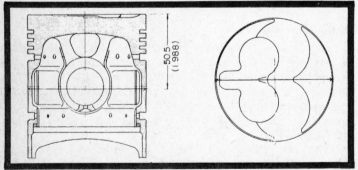

Fig. 7-31. Piston-diameter measuring points. (Courtesy Marine Engine Div., Chrysler Corp.)

162

## Installing Piston Pins

Mike the piston pins. Depending upon your shop standards and the history of the engine, it may well be a good idea to have them Magnifluxed. The latter step would be almost mandatory if the engine had suffered lugging or detonation damage.

Align any match marks on the piston and insert the pin. Excessive tightness at the pin bosses means piston or pin damage. Pressed-in pins are not popular with diesel manufacturers, but have been used. Support the piston and press the pin in two stages: Stop at the point of entry to the lower boss, release pressure to allow the piston to take its as-cast shape, and press the pin home. If the pin is driven home in one shot, the lower boss may be shaved. Full-floating pins should go home with palm pressure in a warm piston.

Install *new* lockrings, being careful not to force them beyond their elastic limits. As mentioned previously, some lockrings can be fitted by hand pressure alone.

## Connecting Rods

Connecting rods are typically I-beam steel forgings. Rods are identified by cylinder (or should be) and have a definite right and left as defined by oil ports. The caps are not interchangeable and are aligned by means of match marks. Reversing the cap or mixing caps from different rods will result in early and dramatic engine failure.

End journals should be miked at two places to detect possible taper (Fig. 7-32). Subtract the bearing diameter from the wristpin and journal diameters to determine the running

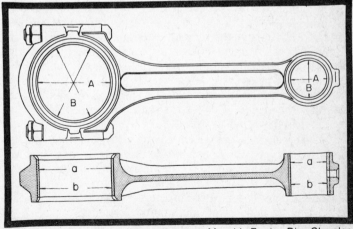

Fig. 7-32. Rod measuring points. (Courtesy Manside Engine Div., Chrysler Corp.)

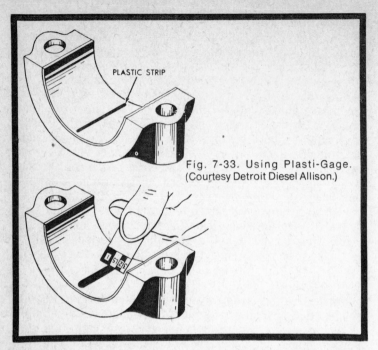

PLASTIC STRIP

Fig. 7-33. Using Plasti-Gage.
(Courtesy Detroit Diesel Allison.)

clearance. This most critical dimension varies between engines but should not exceed 0.004 in. or be less than 0.002 in. Piston pins should have no more than 0.001 in. clearance at the small-end bushing.

Plasti-Gage, a product of the Perfect Circle Corporation, is often used to measure big-end clearances. The advantage of this method is its directness and simplicity. The engine can be left in place while the rod (and main) bearings are checked from below. Remove the rod cap, wipe all traces of oil off the inserts, and lay a length of the plastic strip along the bearing. Generally it is applied on the axis of the bore (the part of the journal which takes the severest wear). Install the rod cap and torque to specs—without turning the shaft. Drop the cap and measure the width of the strip with the indicator on the package, as shown in Fig. 7-33.

Insert bearings are replaced in pairs and may not be identical. That is, the upper and lower halves may have oil ports and grooves which have to be matched to the rod and cap. These bearings are available in a number of grades, depending upon the severity of the service. Diesels survive best with the copper- or aluminum-overplated types. It is important that the shells fit the rod bores tightly. To insure this, dimension A in Fig. 7-34 is larger than dimension B. The

164

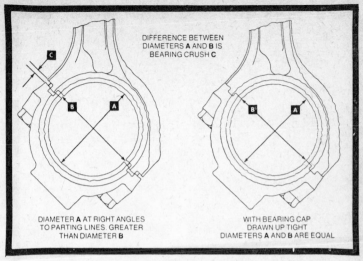

DIFFERENCE BETWEEN DIAMETERS **A** AND **B** IS BEARING CRUSH **C**

DIAMETER **A** AT RIGHT ANGLES TO PARTING LINES. GREATER THAN DIAMETER **B**

WITH BEARING CAP DRAWN UP TIGHT DIAMETERS **A** AND **B** ARE EQUAL

Fig. 7-34. Bearing crush. (Courtesy International Harvester.)

difference is known as the bearing crush. When the bearing is drawn up, the inserts are compressed, equalizing the diameter.

The rods should be placed in a precision fixture (available from the engine manufacturer) and checked for bending and twisting. Piston wear patterns may alert you to the problem. A bent rod (Fig. 7-35) gives an hourglass-shaped wear pattern on

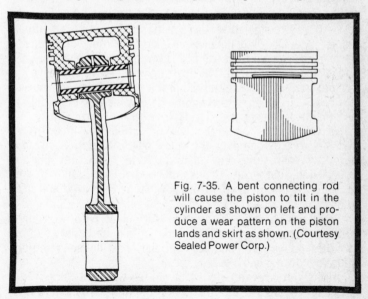

Fig. 7-35. A bent connecting rod will cause the piston to tilt in the cylinder as shown on left and produce a wear pattern on the piston lands and skirt as shown. (Courtesy Sealed Power Corp.)

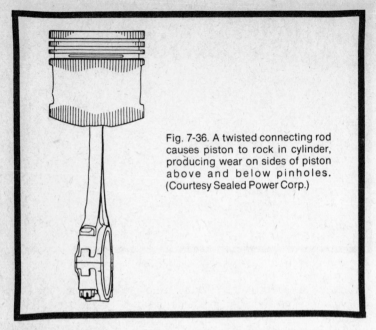

Fig. 7-36. A twisted connecting rod causes piston to rock in cylinder, producing wear on sides of piston above and below pinholes. (Courtesy Sealed Power Corp.)

the piston skirt. A twisted rod causes the piston to rock, localizing contact at the base of the skirt and below the ring belt (Fig. 7-36). In some instances rods can be successfully straightened without unduly weakening them.

Rods should be Magnifluxed. Interpretation of the crack structure thus revealed requires some finesse. Any rod which has seen service will develop cracks. One must distinguish between inconsequential surface flaws and cracks which can lead to structural failure. The drawing in Fig. 7-37 can serve as a guide. In general, longitudinal cracks are not serious unless they are $1/32$ in. deep, which can be determined by grinding at the center of the crack. Transverse cracks are causes of concern since they can be the first indication of fatigue. If the cracks do not extend over the edges of the H-section, are no more than ½ in. long, and less than $1/64$ in. deep, they can be ground and feathered. Cracks over the H-section can be removed if 0.005 in. deep or less. Cracks in the small end are grounds for rejection.

Small-end bushings are pressed into place with reference to the oil port and are finish-reamed. Loose bushings are a sign that the rod has overheated, and they can cause a major failure by turning and blocking the oil port.

After you are satisfied that all is well with the rod, assemble it—aligning the match marks—and the piston to the

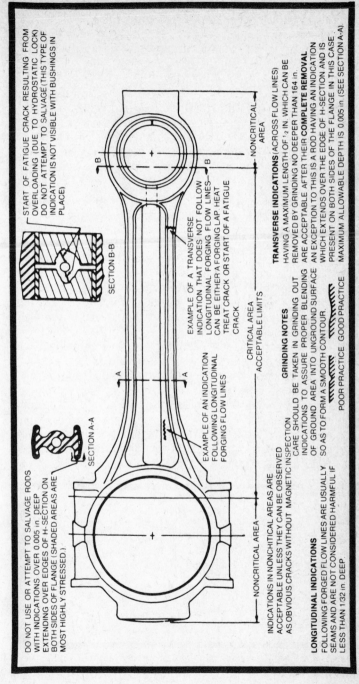

DO NOT USE OR ATTEMPT TO SALVAGE RODS WITH INDICATIONS OVER 0.005 in. DEEP EXTENDING OVER EDGES OF H-SECTION ON BOTH SIDES OF FLANGE (SHADED AREAS ARE MOST HIGHLY STRESSED.)

START OF FATIGUE CRACK RESULTING FROM OVERLOADING (DUE TO HYDROSTATIC LOCK) DO NOT ATTEMPT TO SALVAGE (THIS TYPE OF INDICATION IS NOT VISIBLE WITH BUSHINGS IN PLACE)

NONCRITICAL AREA

SECTION B-B

EXAMPLE OF A TRANSVERSE INDICATION THAT DOES NOT FOLLOW LONGITUDINAL FORGING FLOW LINES. CAN BE EITHER A FORGING LAP HEAT TREAT CRACK OR START OF A FATIGUE CRACK

CRITICAL AREA ACCEPTABLE LIMITS

SECTION A-A

EXAMPLE OF AN INDICATION FOLLOWING LONGITUDINAL FORGING FLOW LINES

**GRINDING NOTES**
CARE SHOULD BE TAKEN IN GRINDING OUT INDICATIONS TO ASSURE PROPER BLENDING OF GROUND AREA INTO UNGROUND SURFACE SO AS TO FORM A SMOOTH CONTOUR.
〰〰〰 POOR PRACTICE   〰〰〰 GOOD PRACTICE

**TRANSVERSE INDICATIONS** (ACROSS FLOW LINES) HAVING A MAXIMUM LENGTH OF 1/2 in. WHICH CAN BE REMOVED BY GRINDING NO DEEPER THAN 1/64 in. ARE ACCEPTABLE AFTER THEIR **COMPLETE REMOVAL** AN EXCEPTION TO THIS IS A ROD HAVING AN INDICATION WHICH EXTENDS OVER THE EDGE OF H-SECTION AND IS PRESENT ON BOTH SIDES OF THE FLANGE IN THIS CASE. MAXIMUM ALLOWABLE DEPTH IS 0.005 in. (SEE SECTION A-A).

NONCRITICAL AREA

INDICATIONS IN NONCRITICAL AREAS ARE ACCEPTABLE UNLESS THEY CAN BE OBSERVED AS OBVIOUS CRACKS WITHOUT MAGNETIC INSPECTION

**LONGITUDINAL INDICATIONS**
FOLLOWING FORGED FLOW LINES ARE USUALLY SEAMS AND ARE NOT CONSIDERED HARMFUL IF LESS THAN 1/32 in. DEEP.

Fig 7-37. Interpretating Magnaflux indications. (Courtesy Detroit Diesel Allison.)

167

crankshaft. Use plenty of clean engine lube and tighten the capscrews in small increments, keeping pressure on each side of the inserts about equal. Otherwise the bearing may distort. With an approved lubricant on the capscrew threads, torque to specifications.

## CRANKSHAFTS

Diesel crankshafts are for the most part steel forgings. Some of the smaller engines use iron shafts, which may be shot-peened for fatigue resistance. Remove the shaft from the block, blow out, or (better) remove the plugs and with a wire brush clean the drilled passages. Make this series of inspections:

- Examine the journal finish for small scratches and light ridging caused by the oil grooves in the bearings. These should be removed with crocus cloth. Use fuel oil for a lubricant and check your progress with a micrometer. It is easy to turn the crank out of round in this operation.
- Check the oil seal contact points at either end of the shaft. Remove small imperfections with crocus cloth.
- Thrust faces should be checked for wear. If the imperfections are too severe to remedy with a stone, have the crankshaft machined.
- Check the timing-gear teeth for wear and chipping. The gear can be removed with a suitable puller; should replacement be necessary, you should replace the other gears in the train.
- Check crankshaft trueness with V-blocks and a dial indicator.
- Mike the crankshaft journals and pins as shown in Fig. 7-38. Compare taper, out-of-roundness, and diameter with factory-specified limits.

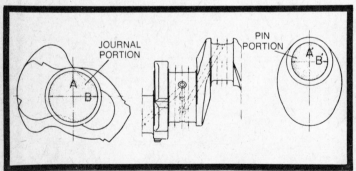

Fig. 7-38. Crankshaft measuring points. (Courtesy Marine Engine Div., Chrysler Corp.)

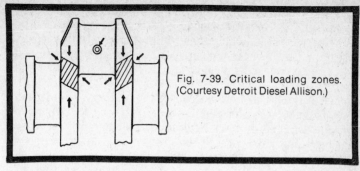

Fig. 7-39. Critical loading zones. (Courtesy Detroit Diesel Allison.)

## Flaw Testing

Flaw testing is generally done with the Magniflux process, although some shops prefer to use the fluorescent-particle method. Both function on the principle that cracks in the surface of the crankshaft take on magnetic polarity when the crank is put in a magnetic field. Iron particles adhere to the edges of these cracks, making them visible. The fluorescent particle method is particularly sensitive since the metal particles fluoresce and glow under black light.

Most cracks are of little concern since the shaft is loaded only at the points indicated in Fig. 7-39. The strength of the shaft is impaired by crack formations which follow these stresses, as shown in Fig. 7-40. These cracks radiate out a 45° to the crank centerline and will eventually result in a complete break.

When you find a failed crankshaft, it is only natural to try to determine the cause of failure. Crankshafts are flaw-tested at the factory and can be assumed to be good when new. Bending stresses are caused by unbalanced forces acting on the pistons (the pistons have inertia, and a reversal of direction occurs twice each revolution) and by the pressure developed in the power stroke. There is little that a mechanic can do to counteract these forces other than to calibrate the

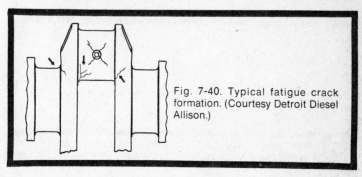

Fig. 7-40. Typical fatigue crack formation. (Courtesy Detroit Diesel Allison.)

injectors carefully and match the weight of the piston assemblies. Abnormal bending forces are generated by main-bearing bore misalignment, improperly fitted bearings, loose main-bearing caps, unbalanced pulleys, or overtightened belts. Cracks caused by bending start at the crankpin fillet and progress diagonally across.

The distribution of cracks caused by torsional, or twisting, forces is the same as for bending forces. All crankshafts have a natural period of torsional vibration, which is influenced by the length/diameter ratio of the crank, the overlap between crankpins and main journals, and the kind of material used. Engineers are careful to design the crankshaft so that its natural periodicity occurs at a much higher speed than the engine is capable of turning. However, a loose flywheel or vibration damper can cause the crank to wind and unwind like a giant spring. Unusual loads, especially when felt in conjunction with a maladjusted governor, can also cause torsional damage.

**Crankshaft Grinding**

Bearings are available in small oversizes (0.001 and 0.002 in.) to compensate for wear. The first regrind is 0.010 in. Some crankshafts will tolerate as much as 0.040 in., although the heat treatment is endangered at this depth. The crankpin and main-journal fillets deserve special attention. Flat fillets invite trouble, since they act as stress risers. Gently radius the fillets as shown in the left drawing in Fig. 7-41.

All journals and pins should be ground, even if only one has failed. Use plenty of lubricant to reduce the possibility of burning the journal. Radius the oil holes with a stone and check the crankshaft again for flaws with one of the magnetic particle methods.

The best course of action is to take these superhard shafts to a renewal station and let them do the work. But a capable machinist can handle the job if he observes these precautions:

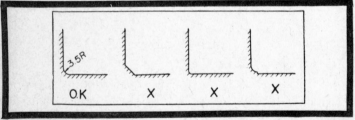

Fig. 7-41. Fillet profiles. The exact radius depends upon the specs. Avoid sharp corners and flat surfaces. (Courtesy Marine Engine Div., Chrysler Corp.)

170

- The wheel must not be too hard. Best results are obtained with an aluminum oxide wheel of $M$ hardness.
- Never attempt to dress the wheel by hand. Use a mechanical or automatic dresser.
- Lubricate with straight cutting oils.
- A good combination is 6500 surface feet per minute, with a work spindle speed of 40—45 rpm.
- Feed rates should be slower than normal.
- Check the dimensions as you would with ordinary crankshafts, but allow plenty of time for the shaft to cool.
- If possible, Magniflux the shaft, since grinding is quite likely to trigger surface flaws.

Since some shafts are suceptible to heat damage, it is good practice to check the hardness before lapping. Perhaps the quickest and surest way to do this is to use Tarasov etch. Clean the shaft with scouring powder or a good commercial solvent. Wash thoroughly and rinse with alcohol. Apply etchant No. 1 (a solution of 4 parts nitric acid in 96 parts water). It is important to pour the acid into the water, not vice versa.

Rinse with clean water and dry. If you use compressed air, see that the system filter traps are clean. Apply etchant No. 2, which consists of 2 parts hydrochloric acid in 98 parts acetone. Acetone is highly flammable and has a sharp odor which may produce dizziness or other unpleasant reactions when used in unventilated areas, so allow yourself plenty of breathing room.

The shaft will go through a color change if it has been burned. Areas that have been hardened by excessive heat will appear white; annealed areas turn black or dark gray. Unaffected areas are neutral gray. If any color other than gray is present, the shaft should be scrapped and the machinist should try again, this time with a softer wheel, a slower feed rate, or a higher work spindle speed. Some experimentation may be necessary to find a combination which works.

## MAIN BEARINGS

The main-bearing caps are usually marked *1, 2, 3,* etc. and have a definite right or left relative to the camshaft. If they are not marked, scribe them before disassembly. One main doubles as the thrust bearing. Check fore-and-aft crankshaft play with a bar and feeler gage or dial indicator prior to disassembly. A thicker bearing or additional shims may be needed.

A few main-bearing caps are assembled with shims, which simplifies adjustment. However, do not attempt to shim caps which have not been so equipped at the factory. Upper and lower insert shells are normally identified by an oil hole. Reversal of the inserts will block the lubricant flow to the bearing.

It is always good practice to check the trueness of the block. There are several ways to do this, most of them requiring rather elaborate fixtures and factory gages. An easy way is to use the engine's own crankshaft after having it checked and straightened as needed. Smear the crank journals with prussian blue and lower the crank carefully onto the bearings. Torque the caps to specifications and turn the crankshaft one-half revolution. Lift the crankshaft carefully and inspect the upper bearing shells for an even transfer of bluing. A misaligned bearing will show only partial contact.

Bearing bosses can be reworked with the appropriate reamer and fitted with inserts to compensate for the material removal. This may be the only route to take for some engines. Some engine makers supply replacement caps which have a surplus of metal at the parting surface. These caps can be jiggled to fit, although the machining is not a job you would try to do during the lunch hour.

## OIL SEALS

Oil seals can be a headache, especially if one takes a notion to leak just after an overhaul. Premature failure is almost always the mechanic's fault and can be traced to improper installation.

Rope or strip seals are still used at the aft end of the crankshaft on some engines. These seals depend upon the resilience of the material for wiping action and so must be installed with the proper amount of compression. Remove the old seal from the grooves and install a new one with thumb pressure. (You may want to coat the groove—but not the seal face—with stickum.) The seal should not be twisted or locally bound in the groove.

Now comes the critical part. Obtain a mandrel of bearing boss diameter (journal plus twice the thickness of the inserts) and, with a soft-faced mallet, drive the sewl home. Using this tool as a ram, you can cut the ends of the seal flush with the bearing parting line. Without one of these tools you will have to leave some of the seal protruding above the parting line in the elusive hope that the seal will compact as the caps are tightened. Undoubtedly some compaction does take place, but the seal ends turn down and becomes trapped between the bearing and cap, increasing the running clearance.

Lubricate the seal with a grease containing molybdenum disulfide to assist in break-in.

Synthetic seals, usually made of Neoprene, are used on all full-circle applications and on the power takeoff end of many crankshafts as well. These seals work in a manner analogous to piston rings. They are preloaded to bear against the shaft and designed so that oil pressure on the wet side increases the force of contact.

These seals must be installed with the proper tools. If the seal must pass over a keyway, obtain a seal protector (a thin tube which slides over the shaft) or at least cover the keyway with masking tape. Drive the seal into place with a bar of the correct diameter. The numbered side is the driven side in most applications. The steep side of the lip profile is the wet side. It is good practice to use a nonhardening sealant on the back of metallic seals. Plastic coated seals are intended to conform to irregularities in the bore and do not need sealant.

More elaborate seals require special one-of-a-kind factory tools. The better engines often incorporate wear sleeves over the shafts, either as original equipment or as a field option. Figure 7-42 shows the use of a wear sleeve on a Detroit Diesel crankshaft. The sleeve makes an interference fit over the shaft and is further secured with a coating of shellac. Worn sleeves are cut off with a chisel or peened to stretch the metal. The latter method is preferred since there is less chance of damaging the shaft.

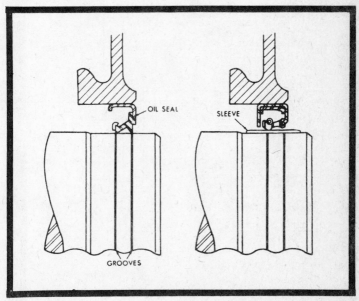

Fig. 7-42. Wear sleeve installation. (Courtesy GM Detroit Diesel Allison.)

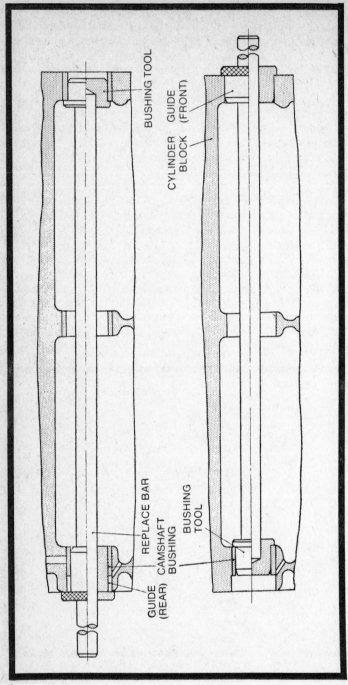

BUSHING TOOL

CYLINDER GUIDE (FRONT)

CYLINDER BLOCK

REPLACE BAR

CAMSHAFT BUSHING

BUSHING TOOL

GUIDE (REAR)

Fig. 7-43. Camshaft bearing. (Courtesy Marine Engine Div., Chrysler Corp.)

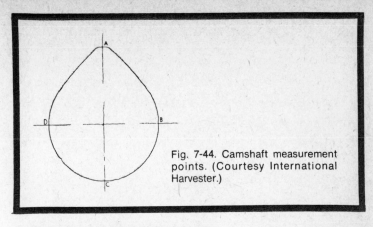

Fig. 7-44. Camshaft measurement points. (Courtesy International Harvester.)

## CAMSHAFTS AND DRIVE GEARS

Inspect the accessory drive gear train for tooth damage and lash (easiest done by meshing a piece of solder between the gears). Crankshaft and camshaft gears are pressed into place and further secured by pins, keys, and bolts. Gears can be pulled or, in the case of a few, split at the pin with a chisel. Check for general fit.

The cam and balance shafts are supported on bushings. Combination puller/drivers are used to remove and install them. Figure 7-43 illustrates one removal sequence. The tool is available from auto parts jobbers or the engine maker. Note the position of any identifying marks on the bushings. On GM engines, for example, the notch in the bushing must be down to align the oil ports.

Subtract measurement $BD$ from $AC$ (Fig. 7-44) to determine the lift. Compare this figure with the builder's specifications to determine the amount of cam wear. Normally camshafts outlast every frictional surface in the engine. However, wear accelerates rapidly once the case hardening has been broken. Wear patterns should be concentrated near the middle of each lobe. Wear on one side means worn tappets and, unless corrected, will lead to early lobe failure.

# Air System  8

We think of an engine in terms of the fuel it burns, because fuel costs money. But an internal combustion engine uses much more air than fuel. Diesel engines have a particularly voracious appetite for air. A naturally aspirated 4-cycle consumes about 200 cu ft of air per pound of fuel; a 2-cycle requires more, and supercharging can double or triple the base figure. Of course, only a small percentage of the air is actually burned in the cylinders. The bulk of it is excess air which cycles through the engine without undergoing any chemical change.

The tremendous amount of air required for diesel operation underscores the importance of air filtration. Airborne particles are sized in microns (1 micron = 0.000039 in.). According to research reported in the SAE *Journal*, the most lethal particles are about 20 microns in diameter. Ten-micron particles are nearly as bad, but dust as finely divided as 4 microns has little measurable effect on engine parts.

Ingested dust may escape out the exhaust valve without doing any damage whatsoever. But those particles which get trapped in the cylinders attack the rings and eventually collect in the oil sump, where they do their nefarious work on other friction surfaces. Once an engine is really dirtied, it is impossible to clean. Abrasives imbed themselves in the bearing metal and are impervious to oil changes and attempts at flushing.

## AIR FILTERS

Two basic filter types are used: *impingement* and *sieve*. The former filters by trapping dust particles in a maze, which

177

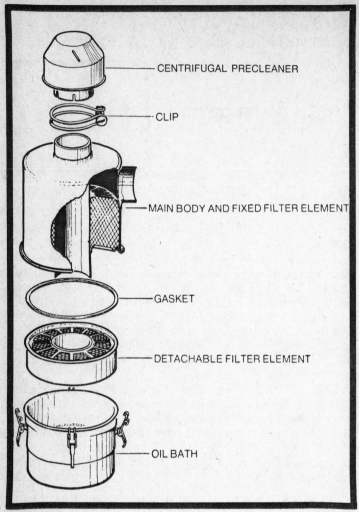

CENTRIFUGAL PRECLEANER

CLIP

MAIN BODY AND FIXED FILTER ELEMENT

GASKET

DETACHABLE FILTER ELEMENT

OIL BATH

Fig. 8-1. Heavy-duty air cleaner, exploded view. (Courtesy GM Bedford Diesel.)

may be steel mesh or polyurethane foam. Oil bath filters are also in this category but employ centrifugal separation. The airstream must change directions several times through the filter. Dust particles tend to be slung out and fall into the oil reservoir. As an additional refinement, the Bedford filter shown in Fig. 8-1 features a *precleaner*. Air entering the precleaner is given a swirl by means of a set of vanes. Dust—as much as 80% of the total—is trapped on the precleaner walls.

Oil bath filters should be serviced by dumping the oil, wiping the sheet metal with a lintless rag, and soaking the wire mesh element in kerosene or other solvent. Dry the mesh with compressed air and fill the reservoir to the prescribed level (marked on the side) with lube. Locate the gasket on the flange and assemble.

*Note*: Overfilling the reservoir or failure to dry the mesh can cause engine runaway as the engine pulls out of the filter.

Polyurethane filters are becoming popular in some applications. The Peugeot filter shown in Fig. 8-2 is typical. The foam is supported by a steel frame. Dust particles are trapped as they make a zigzag passage through the foam. These filters must not be allowed to run dry.

Paper filters (Fig. 8-3) are the most efficient in terms of short-range protection. Capacity is limited by the surface area, but once the filter is broken in with a fine coating of dust, it will stop particles in the range of 1 micron. No other filter does as well. On the other hand, these filters are not cleanable. Rapping the element on a hard surface and reverse blowing helps, but eventually the sieve clogs and the element must be replaced. The interval depends upon the service but should not exceed 300 hours. Do not attempt to clean paper elements with water or solvent. The fibers expand when wetted and render the element airtight.

## TURBOCHARGERS

Developed for high-altitude-aircraft power plants, turbochargers represent some of the most sophisticated aspects of diesel technology (Fig. 8-4). The intriguing aspect of turbocharging is that the power is "free." That is, the compressor is driven by the force of expanding exhaust gases, which is force that would otherwise be lost to the atmosphere. (Nearly a third of the energy released from the fuel of a conventionally aspirated engine goes out the tailpipe in the form of heat and sound.)

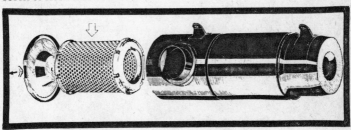

Fig. 8-2. Foam filter. (Courtesy Peugeot.)

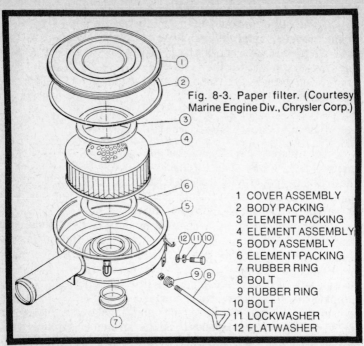

Fig. 8-3. Paper filter. (Courtesy Marine Engine Div., Chrysler Corp.)

1 COVER ASSEMBLY
2 BODY PACKING
3 ELEMENT PACKING
4 ELEMENT ASSEMBLY
5 BODY ASSEMBLY
6 ELEMENT PACKING
7 RUBBER RING
8 BOLT
9 RUBBER RING
10 BOLT
11 LOCKWASHER
12 FLATWASHER

Fig. 8-4. A turbocharged engine—the International DT-466.

The superior power output of turbocharged engines is apparent from a cursory reading of the specifications. The International Harvester DT-414 produces 220 maximum horsepower at 3000 rpm. The same engine without the turbocharger develops only 157 hp. In this instance the gain is 40%, without a significant weight or maintenance penalty. Fuel consumption, when measured on the basis of pounds of fuel per brake-horsepower-hours, drops markedly.

The superior fuel efficiency of turbocharged engines is a function not so much of improved thermal efficiency as of improvements in fuel distribution. In some cases compression

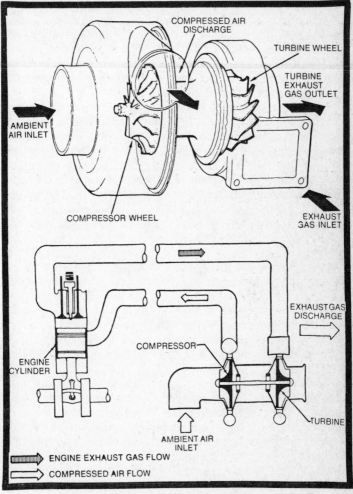

Fig. 8-5. Turbocharger. (Courtesy General Motors Corp.)

ratio must be lowered with a turbocharger. In addition, these engines are less fussy about the quality of fuel they burn, although builders' specifications may not take note of this fact.

In absolute terms, the improved power output is a result of the extra air used to cool valves and piston crowns, enabling the engine to burn a heavier charge of fuel in each cycle without reliability suffering. The turbocharger is better thought of as an adjunct to the cooling system than as an air pump.

But this does not mean that any engine can or should be fitted with a turbocharger. The additional air density may spell trouble for parts which have not been stressed for what is the equivalent of a higher compression ratio. Factory-built turbocharged engines have many detail modifications which may not be obvious on first glance. The head, piston crown, skirt clearances, head gasket, valve timing, and oil system may be altered. International turbocharged engines, for example, cool the piston with oil jets.

Even the most workmanlike factory installation is not free of certain handicaps. In the first place the turbocharger adds cost to the engine. The Daytona Marine 500D lists at $4500; turbocharging adds $1520. Part of this cost is reflected in the detailing, and the rest is the inescapable penalty connected with exotic alloys and turbine wheels which spin at 25,000 rpm. There is no cheap way out on one of these devices. The weight penalty depends upon the application. In some instances the designers were able to forego the penalty entirely. But typically the turbocharger and related hardware adds about a hundred pounds.

A more serious difficulty is the space which is filled by the charger and related plumbing. Exhaust pipes must be routed to the turbine in gentle, large-radius curves. Further space is required if the designer opts for an *intercooler*. An intercooler is a heat exchanger connected in series with the turbocharger output. Air heats under compression, and by bleeding some of this heat into the coolant, a denser charge can be packed into the cylinders.

A turbocharger is a simple machine, consisting of a compressor wheel and a turbine wheel turning together on a common shaft (Fig. 8-5). The turbine wheel must be made of high-temperature alloys equivalent to that used in jet aircraft engines. Both must be balanced to rigid standards beyond that achieved by ordinary manufacturing techniques. Lubricating oil for the shaft bearings comes by way of a flexible line from the engine lube pump output and leaves through a dump line. Because of the rotating speed and the extreme heat, even a momentary interruption of the oil flow caused by air in the line can be disastrous.

Most turbochargers are as shown—without diffuser vanes. At compression ratios of two atmospheres and above, fixed vanes at the outlet side reduce turbulence and increase pumping efficiency. At lower pressures the same vanes act as an obstruction, increasing the tendency of the 'charger to be "peaky." This characteristic, along with a certain lag in response, is typical of all centrifugal pumps. At low rotational speeds air escapes between the vanes and the impeller casing; as speeds increase the pump becomes more efficient, and output jumps. The lag, or response delay to engine loads, is the result of the inertia of the wheels and shaft. It takes time for them to accelerate.

## Installation

Care must be taken with refilling oil bath air cleaners. Excessive oil will be pulled over by the turbocharger and can lead to loss of power along with possible detonation. Long, unsupported exhaust stacks must not be used without consultation with factory engineers. A turbocharger is a precision device, designed for one job; and so the installation criteria are somewhat inflexible. The exhaust stack should be capped to prevent water from entering the turbocharger during shutdown.

General engine maintenance becomes more exacting. Pay particular attention to the crankcase breather: All things equal, it will tend to clog sooner because of increased blowby due to the higher effective compression ratio. The turbine itself should be capped when taken out of service to prevent the entry of dust and foreign matter.

Inspect the turbocharger periodically. Vibration, which may be obvious or which can be inferred from the difficulty of keeping the mounts tight, means blade or shaft damage. Noise can be traced to the bearings. In either case the turbocharger must be rebuilt.

All ducting and gaskets must be air tight. Failure of the air duct upwind of the compressor will allow dust to enter and erode the vanes. Leaks downwind will rob the engine of power. Inspect the oil lines for leaks and restrictions which would cut oil flow to the bearings. Seal failure will result in high oil consumption, blue smoke, and carbon deposits in the intake manifold. In several cases, the seals can go completely, dumping oil at pump pressure into the cylinders.

Remove the inlet duct and compressor housing (Fig. 8-6, No. 2) to check the blades for eroded or feathered edges and to

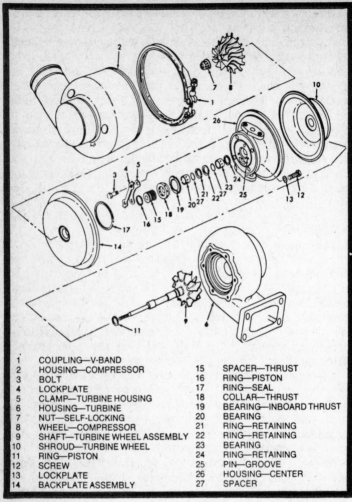

| | | | |
|---|---|---|---|
| 1 | COUPLING—V-BAND | | |
| 2 | HOUSING—COMPRESSOR | 15 | SPACER—THRUST |
| 3 | BOLT | 16 | RING—PISTON |
| 4 | LOCKPLATE | 17 | RING—SEAL |
| 5 | CLAMP—TURBINE HOUSING | 18 | COLLAR—THRUST |
| 6 | HOUSING—TURBINE | 19 | BEARING—INBOARD THRUST |
| 7 | NUT—SELF-LOCKING | 20 | BEARING |
| 8 | WHEEL—COMPRESSOR | 21 | RING—RETAINING |
| 9 | SHAFT—TURBINE WHEEL ASSEMBLY | 22 | RING—RETAINING |
| 10 | SHROUD—TURBINE WHEEL | 23 | BEARING |
| 11 | RING—PISTON | 24 | RING—RETAINING |
| 12 | SCREW | 25 | PIN—GROOVE |
| 13 | LOCKPLATE | 26 | HOUSING—CENTER |
| 14 | BACKPLATE ASSEMBLY | 27 | SPACER |

Fig. 8-6. Turbocharger in exploded view. (Courtesy Detroit Diesel Allison.)

remove dirt and soft carbon buildup. It is not necessary to dismantle the unit further for these checks. Turn the impeller by hand to detect roughness and possible binding in the bearings. End play should typically be 0.005 in. But the spec varies—check the engine builder's recommendations. As a final part of the inspection, allow the engine to reach operating temperature and shut it down, observing the turbine. It should coast to a stop. If it brakes suddenly or loses rpm in fits and snatches, you can assume that there is a bearing or end clearance problem.

If the turbocharger is to be fitted on a new or overhauled virgin engine, it is good practice to run it for at least one hour prior to installation. Otherwise foreign particles—bearing shavings and the like—might find their way into the turbocharger.

Using a new gasket, secure the unit to the exhaust manifold. Install the oil lines, leaving the oil supply line disconnected at the turbocharger. Pour about 5 oz of clean engine oil into the oil inlet port in the center housing. This will provide oil for initial running. Turn the wheels by hand to distribute the oil. Fill the oil supply line with oil.

Secure the wheels with a socket wrench and start the engine. Do not connect the oil supply line to the center housing until oil spills out of it. After you are satisfied that no air is trapped in the line and that oil pressure is adequate, connect it to the center housing and release the wheel.

## Disassembly

Allow the exhaust plumbing to cool before you undo the fasteners. Otherwise the piping and flange may warp. Clean the exterior of the turbocharger will a good, noncaustic solvent. Some shop solvents will attack aluminum. Using the unit pictured in Fig. 8-6 as a reference, remove the coupling band (1) and compressor housing (2). Remove the turbine housing (6) from the center housing (26). It is secured by a bolt-and-clamp arrangement (3 and 5). Separate the castings with a soft-faced mallet. Exercise extreme care during this operation: Casual handling of the turbine housing can nick the wheel, throwing it off balance.

Further disassembly requires a holding fixture such as the one shown in the working drawings in Fig. 8-7. Dimensions and blade contour apply to turbochargers fitted to Detroit Diesel engines and are by no means universal.

With the compressor wheel (8, Fig. 8-6) secured by the fixture, remove the nut. This operation must be done with circumspection since it is quite easy to bend the shaft in the process. Use a T-handle wrench and a double universal socket to isolate the shaft from side loads. Next, place the center housing in an oven which has been heated to 350°F. Allow the parts to heat for no more than 10 minutes. The compressor wheel should expand enough to slide off the shaft (9). Remove the seal (11) and the back-plate assembly (14), which is secured by screws. Note the sequence of parts in the drawing from 16 to 24.

Clean the parts thoroughly, blow dry, and scrutinize shaft and bearings, turbine vanes, seals, and housings (for signs of

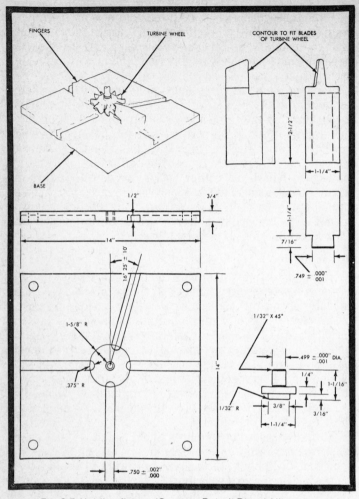

Fig. 8-7. Holding fixture. (Courtesy Detroit Diesel Allison.)

rubbing or galling). If the bearings or thrust washer show any damage at all, they should be replaced as a complete set. Damage includes nicks, abrasions, shellac, and foreign particles.

## Assembly

Almost surgical standards of cleanliness are required during final assembly. Use filtered compressed air, lint-free rags (or better, paper towels), and spotless tools. Prelubricate the bearings (20 and 23) with clean engine oil. Assemble with new retainer rings and piston ring; *do not force the piston ring into place*. The inboard thrust bearing (*19*) aligns with pins (*25*) in the center housing. Lubricate and install the thrust collar along with a new seal ring (*17*), which fits into a groove in the back-plate assembly (*14*). Be sure to align the oil supply holes in the center housing (*26*) and the back-plate assembly. Use new lock plates on the bolts and torque to 83 in.-lb. Take up the fasteners evenly in several stages. Bend over the tangs on the lock plates. Locate a new piston ring (*11*) on the shaft.    With the shroud (*10*) against the center housing, carefully insert the shaft assembly. Do not force the piston ring; gently push and rock until the ring seats and the shaft bottoms. Be careful not to scratch the bearings.

Heat the compressor wheel to 350°F in an oven or in oil, for no longer than 10 minutes, to expand it enough to slip over the shaft. Secure the shaft in the fixture and assemble the compressor wheel. Inspect the nut and wheel faces for scores or other irregularities. Torque to 20 in.-lb. Now comes the critical part: Torque readings are not accurate enough for this fastener. Continue to tighten the nut until the shaft stretches 0.008−0.009 in. Take the precautions described in the previous section to prevent the shaft from bending under side forces.

Before going any further, check the fore-and-aft play of the shaft with a dial indicator. Axial movement should be between 0.004 and 0.007 in. Continue the assembly by securing the turbine housing (*6*) with clamps (*5*), lock plates (*6*), and bolts (*3*). Tighten to 170 in.-lb. and crimp the lock plates to

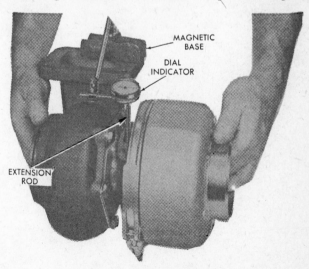

Fig. 8-8. Checking radial play. (Courtesy Detroit Diesel Allison.)

187

prevent the bolts from turning out. Finally, install the compressor housing (2) and band (1). Tighten the band to 35–45 in.-lb.

**Final Bench Check**

Push the shaft-and-wheel assembly to the compressor side as far as possible. Rotate the assembly in this position and check for binds. Do the same for the other extreme of travel.

The next test is a bit tricky. Mount a dial indicator as shown in Fig. 8-8. The extension bar must be centered on the shaft and cannot be in rubbing contact with the sides of the oil drain passage. Lift the shaft upward, applying pressure on both wheels simultaneously. Note the reading. Now push the shaft down in the same manner, away from the indicator. The displacement should be greater than 0.003 in. and less than 0.007 in.

# Electrical Fundamentals 9

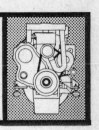

Most small diesel engines are fitted with an electric starter, battery, and generator. The circuit may include glow plugs for cold starting and electrically operated instruments such as pyrometers and flow meters. The diesel technician should have knowledge of electricity.

This knowledge cannot be gleaned from the hardware. Just looking at an alternator will not tell you much about its workings. The only way to become even remotely competent in electrical work is to have some knowledge of basic theory. This chapter is a brief, almost entirely nonmathematical, discussion of the theory.

## ELECTRONS

Atoms are the building blocks of all matter. These atoms are widely distributed: If we enlarged the scale to make atoms the size of pinheads, there would be approximately one atom per cubic yard of nothingness. But as tiny and as few as they are, atoms (or molecules, which are atoms in combination) are responsible for the characteristics of matter. The density of a substance, its chemical stability, thermal and electrical conductivity, color, hardness, and all its other characteristics are fixed by the atomic structure.

The atom is composed of numerous subatomic particles. Using high-energy disintegration techniques, scientists are discovering new particles almost on an annual basis. Some are reverse images of the others; some exist for only a few millionths of a second. But, we are only interested in the relatively gross particles whose behavior has been reasonably well understood for generations.

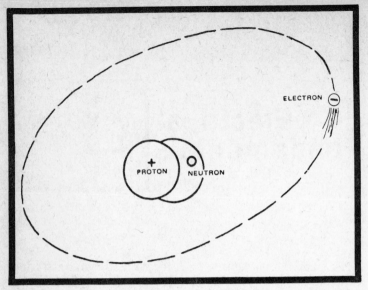

Fig. 9-1. Electron in orbit around a nucleus consisting of a proton and a neutron. (Not drawn to scale.)

In broad terms the atom consists of a nucleus and one or more electrons in orbit around it. The nucleus has at least one positively charged *proton* and may have one or more electrically neutral *neutrons*. The particles make up most of the atom's mass. The orbiting *electrons* have a negative charge. All electrons, insofar as is known, are identical. All have the same electrical potency. Their orbits are balanced by centripetal force and the pull of the positive charge of the nucleus (Fig. 9-1).

A fundamental electrical law is: Unlike charges attract, like charges repel. Thus two electrons, both carrying negative charges, repel each other. Attraction between opposites and the repulsion of likes are the forces that drive electrons through circuits.

## CIRCUITS

The term *circuit* comes from a Latin root meaning *circle*. It is descriptive since the path of electrons through a conductor is always in the form of a closed loop, bringing them back where they started.

A battery or generator has two *terminals*, or posts. One terminal is negative. It has a surplus of electrons available. The other terminal has a relative scarcity of electrons and is positive.

A *circuit* is a pathway between negative and positive terminals. The relative ease with which the electrons move along the circuit is determined by several factors. One of the most important is the nature of the conductor. Some materials are better conductors than others. Silver is at the top of the list, followed by copper, aluminum, iron, and lead. These and other conductors stand in contrast to that class of materials known as insulators: most gases, including air; wood, rubber, and other organic materials; most plastics; mica; and slate are insulators of varying effectiveness.

Another class of materials is known as *semiconductors*. They share the characterics of both conductors and insulators and under the right conditions can act as either. Silicon and germanium semiconductors are used to convert alternating current to direct current at the alternator and to limit current and voltage outputs. We discuss these devices in more detail later in this chapter.

The ability of materials to pass electrons depends upon their atomic structures. Electrons flood into the circuit from the negative pole. They encounter other electrons already in orbit. The newcomers displace the orbiting electrons (like repels like) toward the positive terminal and, in turn, are captured by the positive charge on the nucleus (unlike attracts). This game of musical chairs continues until the circuit is broken or until the voltages on the terminals equalize. Gold, copper, and other conductors give up their captive electrons with a minimum of fuss. Insulators hold their electrons tightly in orbit and resist the flow of current.

You should not have the impression that electrons flow through various substances in a go/no-go manner. All known materials have some reluctance to give up their orbiting electrons. This reluctance is lessened by extreme cold, but never reaches zero. And all insulators "leak" to some degree. A few vagrant electrons will pass through the heaviest, most inert insulation. In most cases the leakage is too small to be significant. But it exists and can be accelerated by moisture saturation and by chemical changes in the insulator.

### Circuit Characteristics

Circuits, from the simplest to the most convoluted, share these characteristics:

- Electrons move from the negative to the positive terminal. Actually the direction of electron movement makes little difference. In fact, for many years it was thought that electrons moved in the opposite direction—from positive to negative.

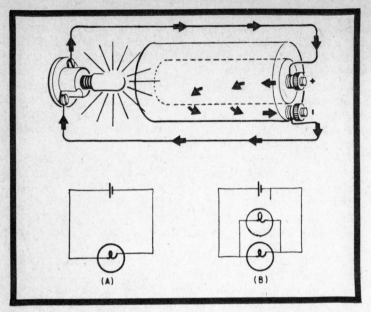

Fig. 9-2. Basic series circuit (top) and its schematic (A). In B, a second lamp has been added with the first.

- The circuit must be complete and unbroken for electrons to flow. An incomplete circuit is described as *open*. The circuit may be opened deliberately by the action of a switch or a fuse, or it might open of its own accord as in the case of a loose connection or a broken wire.

    Since circuits must have some rationale besides the movement of electrons from one pole to the other, they have *loads*, which convert electron movement into heat, light, magnetic flux, or some other useful quantity. These working elements of the circuit are also known as *sinks*. They absorb, or sink, energy and are distinguished from the sources (battery and the generator). A circuit in which the load is bypassed is described as *shorted*. Electrons take the easiest, least resistive path to the positive terminal.

- Circuits may feed single or multiple loads. The arrangement of the loads determines the circuit type.

### Series and Parallel Circuits

A *series* circuit has its loads connected one after the other like beads on a string (Fig. 9-2A). There is only one path for the electrons between the negative and positive terminals. Some readers may remember series Christmas tree lights. Each lamp was rated at 10V; 11 in series required 110V. If any lamp failed, the string went dark.

Pure series circuits are rare today. (They can still be found as filament circuits in some alternating current/direct current radios and in high-voltage applications such as airport runway lights.) But switches, rheostats, relays, fuses, and other circuit controls are necessarily in series with the loads they control.

*Parallel* circuits have the loads arranged like the rungs of a ladder, to provide multiple paths for current (Fig. 9-2B). When loads are connected in parallel, we say they are *shunt*. Circuits associated with diesel engines usually consist of parallel loads and series control elements. The major advantage of the parallel arrangement from a serviceman's point of view is that a single load can open without affecting the other loads. A second advantage is that the voltage remains constant throughout the network.

### Single- and Two-Wire Circuits

However the circuit is arranged—series, parallel, or in some combination of both—it must form a complete path between the positive and negative terminals of the source. This requirement does not mean the conductor must be composed entirely of electrical wire. The engine block, transmission, and mounting frame are not the best conductors from the point of view of their atomic structures. But because of their vast cross-sectional area, these components have almost zero resistance.

In the single-wire system the battery is *grounded*, or, as the British say, *earthed* (Fig. 9-3). These terms seem to have originated from the power station practice of using the earth as a return. With the exception of some Lucas and CAV systems, most modern designs have the negative post grounded. The "hot" cable connects the positive post to the individual loads, which are grounded. Electrons flow from the negative terminal, through the loads, and back to the battery via the wiring.

The single-wire approach combines the virtues of simplicity and economy. But it has drawbacks. Perhaps the most consequential is the tendency for the connections to develop high resistances. One would think that a heavy strap bolted to the block would offer no more resistance than the Happy Hooker. Unfortunately this is not the case. A thin film of oil, rust, or a loose connection is enough to block the flow of current.

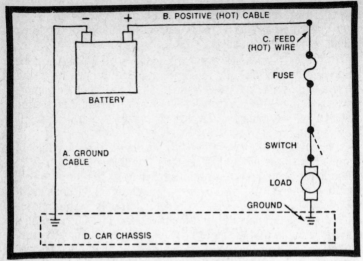

Fig. 9-3. The principle of grounding.

Corrosion problems are made more serious by weather exposure and *electrolysis*. Electrolysis is the same phenomenon which occurs during electroplating. When current passes through two dissimilar metals (e.g., copper and cast iron) in the presence of damp air, one of the metals tends to disintegrate. In the process it undergoes chemical changes which make it a very poor conductor.

Another disadvantage of the single-wire system, and the reason it is not used indiscriminately on aircraft and ocean-going vessels, is the danger of short circuits. Contact between an uninsulated hot wire and the ground will shunt the loads down circuit.

The 2-*wire* system—one wire to the load, and a second wire from it to the positive terminal—is preferred for critical loads and can be used in conjunction with a grounded system. Electrolytic action is almost nil, and shorts occur only in the unlikely event that two bare wires touch.

## ELECTRICAL MEASUREMENTS

*Voltage* is a measure of electrical pressure. In many respects it is analogous to hydraulic pressure. The unit of voltage is the volt, abbreviated *V*. Thus we speak of a 12V or 24V system. Prefixes, keyed to the decimal system, expand the term so we do not need to contend with a series of zeros. The prefix *K* stands for *kilo*, or 1000. A 50 kV powerline delivers 50,000V. At the other end of the scale, *milli* means $1/1000$, and one millivolt (1 mV) is a thousandth of a volt.

The *ampere* shortened sometimes to *amp* and abbreviated A, is a measure of the amount of electrons flowing past a given point in the circuit per second. One ampere represents the flow of $6.25 \times 10^{18}$ electrons per second. Amperage is also referred to as *current intensity* or *quantity*. From the point of view of the loads on the circuit, the amperage is the *draw*. A free-running starter motor may draw 100A, and three times as much under cranking loads.

*Resistance* is measured in units named after G. S. Ohm. Ohms are expressed by the last letter of the greek alphabet, omega ($\Omega$). Thus we may speak of a 200$\Omega$ resistance.

The resistance of a circuit determines the amount of current that flows for a given applied voltage. The resistance depends upon the atomic structure of the conductor—how tightly the electrons are held captive in their orbits—and on certain physical characteristics. The broader the cross-sectional area of the conductor, the less opposition to current. And the longer the path formed by the circuit between the poles of the voltage source, the more resistance. We can think of these two dimensions in terms of ordinary plumbing. The *resistance* to the flow of a liquid in a pipe is inversely related to its diameter (decreases as diameter increases) and directly related to its length. Resistance in the pipe produces heat, exactly as does resistance in an electrical conductor.

Resistance is generally thought of as the electrical equivalent of friction—a kind of excise tax which we must pay to have electron movement. There are, however, positive uses of resistance. Resistive elements can be deliberately introduced in the circuit to reduce current in order to protect delicate components. The heating effect of resistance is used in soldering guns and irons and in the glow plugs employed as starting aids in diesel engines.

## OHM'S LAW

The fundamental law of simple circuits was expressed by the Frenchman G. S. Ohm in the early 1800s in a paper on the effects of heat upon resistance. The law takes three algebraic forms, each based on the following relationship: a potential of 1V drives 1A through a resistance of 1 ohm. In the equations the symbol $E$ stands for *electromotive force*, or, as we now say, volts. An $I$ represents *intensity*, or current, and $R$ is resistance in ohms.

The basic relationship is expressed as

$$I = \frac{E}{R}$$

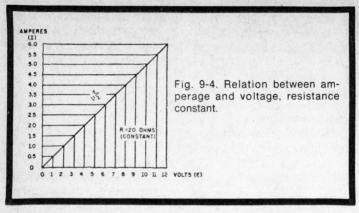

Fig. 9-4. Relation between amperage and voltage, resistance constant.

This form of the equation states that current in amperes equals voltage in volts divided by resistance in ohms. If a circuit with a resistance of 6 ohms is connected across a 12V source, the current is 2A (12/6 = 2). Double the voltage (or halve the resistance), and the current doubles. Figure 9-4 shows this linear (straight-line graph) relationship.

Another form of Ohm's law is

$$R = \frac{E}{I}$$

or resistance equals voltage divided by current. The 12V potential of our hypothetical circuit delivers 2A, which means the resistance is 6 ohms. The relation between amperage and resistance with a constant voltage is illustrated in Fig. 9-5.

Another way of expressing Ohm's law is

$$E = IR$$

or voltage equals current multiplied by resistance. Two amperes through 6 ohms requires a potential of 12V. Double the resistance, and twice the voltage is required to deliver the same amount of current (Fig. 9-6).

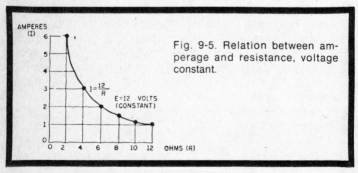

Fig. 9-5. Relation between amperage and resistance, voltage constant.

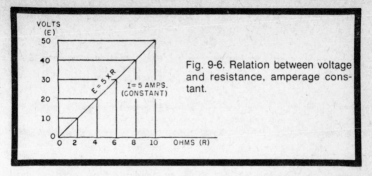

Fig. 9-6. Relation between voltage and resistance, amperage constant.

Various memory aids have been devised to help students remember Ohm's law. One involves an Indian, an eagle, and a rabbit. The Indian ($I$) sees the eagle ($E$) flying over the rabbit ($R$) this gives the relationship $I = E/R$. The eagle sees both the Indian and the rabbit on the same level, or $E = IR$. And the rabbit sees the eagle over the Indian, or $R = E/I$. A visual aid is shown in Fig. 9-7. The circle is divided into three segments. To determine which form of the equation to use, put your finger on the quantity you want to solve for.

All of this may seem academic, and, in truth, few mechanics perform calculations with Ohm's law. It is usually easier to measure all values directly with a meter. Still, it is very important to have an understanding of Ohm's law. It is the best description of simple circuits we have.

For example, suppose the wiring is *shorted*: electrons have found a more direct (shorter) path to the positive terminal of the battery. Ohm's law tells us certain facts about the nature of shorts which we will find useful in troubleshooting. First, a short increases the current in the affected circuit, since current values respond inversely to resistance. This increase will generate heat in the conductor and may even carbonize the insulation. Voltage readings will be low since the short has almost zero resistance. Now suppose we have a partially open circuit caused, say, by a corroded terminal. The current through the terminal will be reduced, which means that the lights or whatever other load is on the

Fig. 9-7. Ohm's law: Cover the unknown with your finger to determine which of the three equations to use.

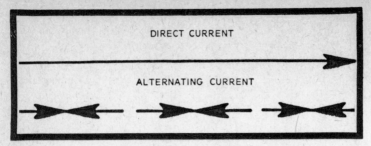

Fig. 9-8. Direct and alternating current.

circuit will operate at less than peak output. The terminal will be warm to the touch, since current is transformed to heat by the presence of resistance. Voltage readings from the terminal to ground will be high on the source side of the resistance and lower than normal past the bottleneck.

## DIRECT AND ALTERNATING CURRENT

Diesel engines may employ direct or alternating currents. The action of *direct*, or *unidirectional*, *current* may be visualized with the aid of the top drawing in Fig. 9-8. *Alternating current* (alternating current) is expressed by the opposed arrows in the lower drawing. Flow is from the negative to the positive poles of the voltage source, but the poles exchange identities, causing the current reversal. The positive becomes negative, and the negative becomes positive.

The graph in Fig. 9-9 represents the rise, fall, and reversal of alternating current. Since the amplitude changes over time, alternating voltage and current values require some

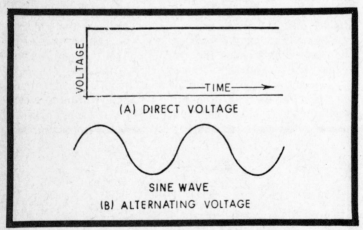

Fig. 9-9. Direct- and alternating-current waveforms.

qualification. The next drawing illustrates the three values most often used.

*Peak-to-peak* values refer to the maximum amplitude of the voltage and amperage outputs in both directions. In Fig. 9-10 the peak-to-peak value is $200V_{p-p}$ (or $200A_{p-p}$). The half-cycle (alternation) on top of the zero reference line is considered to be positive; below the line alternation is negative.

The *average*, or *mean*, value represents an average of all readings. In Fig. 9-10 it is 63.7 units.

The *root mean square* (rms) value is sometimes known as the *effective* value. One rms ampere has the same potential for work as 1A direct current. Unless otherwise specified, alternating current values are *rms* (effective) values and are directly comparable to direct current in terms of the work they can do. Standard meters are equipped with appropriate scales to give rms readings.

The illustration depicts one complete cycle of alternating current. The number of cycles completed per second is the frequency of the current. One cycle per second is the same as one *hertz* (1 Hz). Domestic household current is generated at 60 Hz. The alternating current generators used with diesel engines are variable-frequency devices since they are driven by the engine crankshaft. At high speed a typical diesel alternating current generator will produce 500−600 Hz.

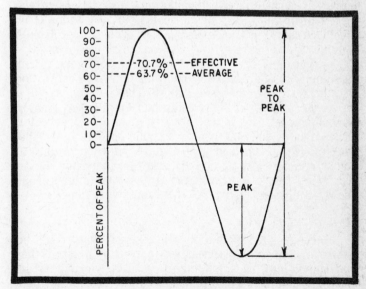

Fig. 9-10. Various valves used to indicate sine-wave amplitude.

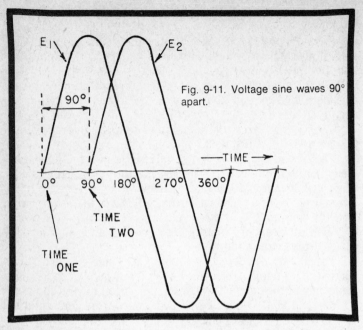

Fig. 9-11. Voltage sine waves 90° apart.

In addition to these special characteristics, alternating current outputs can be superimposed upon each other (see Fig. 9-11). The horizontal axis ($X$-axis) represents zero output—it is the crossover point where alternations reverse direction. In this particular alternator 360° of rotation of the armature represents a complete output cycle from maximum positive to maximum negative and back to maximum positive. The $X$-axis may be calibrated in degrees or units of time (assuming that the alternator operates at a fixed rpm). Note that there are two sine waves shown. These waves are out of phase; $E_1$ leads $E_2$ by 90° of alternator rotation. Alternators used on diesel installations usually generate three output waves 120° apart. Multiphase outputs are smoother than single-phase outputs and give engineers the opportunity to build multiple charging circuits into the alternator. Circuit redundancy gives some insurance against total failure; should one charging loop fail, the two others continue to function.

## MAGNETISM

Electricity and magnetism are distinct phenomena, but related in the sense that one can be converted into the other. Magnetism may be employed to generate electricity, and electricity can be used to produce motion through magnetic attraction and repulsion.

The earth is a giant magnet with magnetic poles located near the geographic poles (Fig. 9-12). Magnetic *fields* extend between the poles over the surface of the earth and through its core. The field consists of *lines of force*, or *flux*. These lines of force have certain characteristics which, although they do not fully explain the phenomenon, at least allow us to predict its behavior. The lines are said to move from the north to the south magnetic pole just as electrons move from a negative to a positive electrical pole. Magnetic lines of force make closed loops, circling around and through the magnet. The poles are the interface between the internal and external paths of the lines of force. When encountering a foreign body, the lines of force tend to stretch and snap back upon themselves like rubber bands. This characteristic is important in the operation of generators.

Lines of force penetrate every known substance, as well as the emptiness of outer space. However, they can be deflected by soft iron. This material attracts and focuses flux in a manner analogous to the action of a lens on light. Lines of force "prefer" to travel through iron, and they digress to take advantage of the *permeability* (magnetic conductivity) of iron (Fig. 9-13).

The permeability of iron is exploited in almost all magnetic machines. Generators, motors, coils, and electromagnets all have iron cores to direct the lines of force most efficiently.

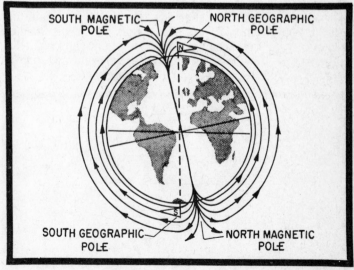

Fig. 9-12. Earth's magnetic field.

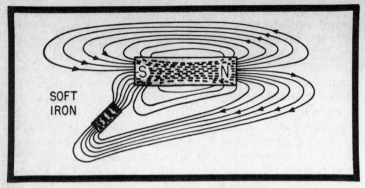

Fig. 9-13. Effects of soft iron on a magnetic field.

When free to pivot, a magnet aligns itself with a magnetic field, as shown in Fig. 9-14. *Unlike magnetic poles attract; like magnetic poles repel.* The south magnetic pole of the small magnet (compass needle) points to the north magnetic pole of the large bar magnet.

## ELECTROMAGNETS

Electromagnets use electricity to produce a magnetic field. Electromagnets are used in starter solenoids, relays, and voltage- and current-limiting devices. The principle upon which these magnets operate was first enunciated by the Danish researcher Hans Christian Oersted. In 1820 he reported a rather puzzling phenomenon: A compass needle, when placed near a conductor, deflected as the circuit was made and broken. Subsequently it was shown that the needle reacted because of the presence of a magnetic field at right angles to the conductor. The field exists as long as current flows, and its intensity is directly proportional to the amperage.

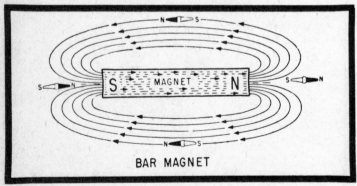

Fig. 9-14. Compass deflections in a magnetic field.

The field around a single strand of conductor is too weak to have practical application. But if the conductor is wound into a coil, the weak fields of the turns reinforce each other (Fig. 9-15). Further reinforcement may be had by inserting an iron bar inside of the coil to give focus to the lines of force. The strength of the electromagnet depends upon the number of turns of the conductor, the length/width ratio of the coil, the current strength, and the permeability of the core material.

## VOLTAGE SOURCES

Starting and charging systems employ two voltage sources. The generator is electromagnetic in nature and is the primary source. The battery, which is electrochemical in nature, is carried primarily for starting. Also, in case of overloads, the battery can send energy into the circuit.

## GENERATOR PRINCIPLES

In 1831 Michael Faraday reported that he had induced electricity in a conductor by means of magnetic action. His apparatus consisted of two coils, wound over each other but not electrically connected. When he made and broke the circuit to one coil, momentary bursts of current were induced in the second. Then Faraday wound the coils over a metal bar and observed a large jump in induced voltage. Finally he reproduced the experiment with a permanent magnet. Moving a conductor across a magnetic field produced voltage. The intensity of the voltage depended upon the strength of the field,

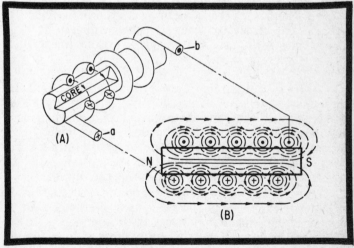

Fig. 9-15. Electromagnet construction showing how magnetic fields mutually reinforce one another.

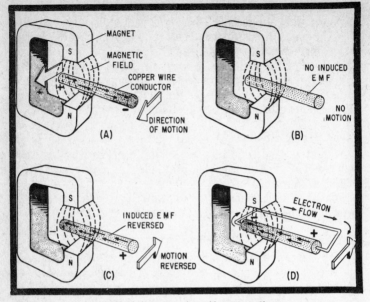

Fig. 9-16. Voltage produced by magnetism.

the rapidity of movement, and the angle of movement. His generator was most efficient when the conductor cut the lines of force at right angles.

The rubber band effect mentioned earlier helps one to visualize what happens when current is induced in the conductor. A secondary magnetic field is set up which resists the movement of the conductor through the primary field. The greater the amount of induced current, the greater the resistance, and the more power required to overcome it.

### Alternator

Figure 9-16 illustrates Faraday's apparatus. Note that there must be relative movement between the magnetic field and conductor. Either may be fixed as long as one can move. Note also that the direction of current is determined by the direction of movement. Of course, this shuttle generator is hardly efficient: The magnet or conductor must be accelerated, stopped, and reversed during each cycle.

The next step was to convert Faraday's laboratory model to rotary motion (Fig. 9-17). The output alternates with the position of the armature relative to the fixed magnets. In Fig. 9-17A the windings are parallel to the lines of force and the output is zero. Ninety degrees later the output reaches its highest value since the armature windings are at right angles

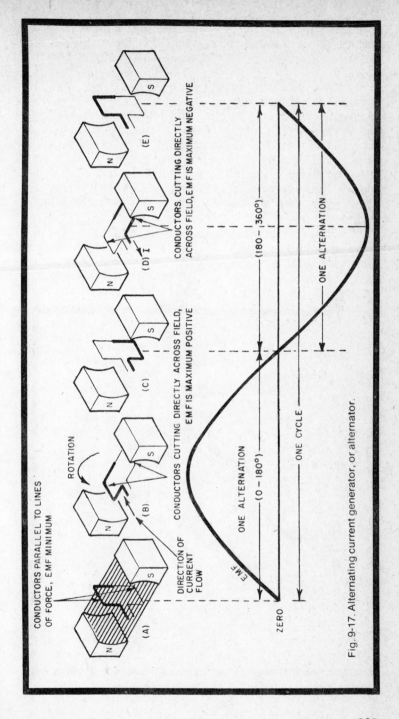

Fig. 9-17. Alternating current generator, or alternator.

205

to the field. At 180° of rotation (*C*) the output is again zero. This position coincides with a polarity shift. For the remainder of the cycle, the movement of the windings relative to the field is reversed.

To determine the direction of current flow from a generator armature, position the left hand so that the thumb points in the direction of current movement and the forefinger points in the direction of magnetic flux (from the north to the south pole). Extend the middle finger 90° from the forefinger; it will point to the direction of current. As convoluted as the left-hand rule for generators sounds, it testifies to the fact that polarity shifts every 180° of armature rotation. This is true of all dynamos, whether the machine is dc generator or ac generator (alternator).

The frequency in hertz depends upon the rotational speed of the armature and upon the number of magnetic poles. Figure 9-18 illustrates a 4-pole (2-magnet) alternator which, for a given rpm, has twice the frequency of the single-pole machine in the previous illustration. Normally, alternators

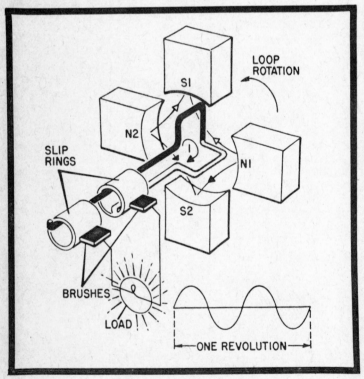

Fig. 9-18. Four-pole alternator.

supplied for diesel engines have four magnetic poles and are designed to peak at 6000 rpm. Their frequency is given by

$$f = \frac{P \times \text{rpm}}{120}$$

where $f$ is frequency in hertz and $P$ is the number of poles. Thus, at 6000 rpm, a typical alternator should deliver current at a frequency of 2000 Hz.

The alternators depicted in Fig. 9-17 and 9-18 are, of course, simplifications. Actual alternators differ from these drawings in two important ways: The fields are not fixed, but rotate as part of an assembly called the *rotor*. Nor are the fields permanent magnets; instead they are electromagnets whose strength is controlled by the voltage regulator. Electrical connection to the rotor is made by a pair of brushes and sliprings.

### DC Generator

Direct-current generators develop an alternating current which is mechanically rectified by the commutator—brush assembly. The commutator is a split copper ring with the segments insulated from each other. The brushes are carbon bars which are spring-loaded to bear against the commutator. The output is pulsating direct current, as shown in Fig. 9-19. The pulse frequency depends upon the number of armature loops, field poles, and rpm. A typical generator has 24—28 armature loops and 4 poles. The fields are electromagnets and

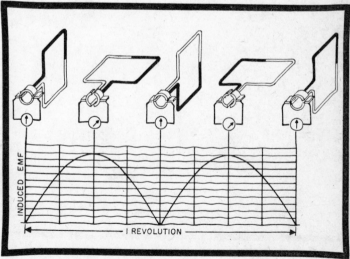

Fig. 9-19. Single-loop direct current generator.

are energized by 8—12% of the generator output current. The exact amount of current detailed for this *excitation* depends upon the regulator, which, in turn, responds to the load and state of charge of the generator. A few surviving generators achieve output regulation by means of a third brush.

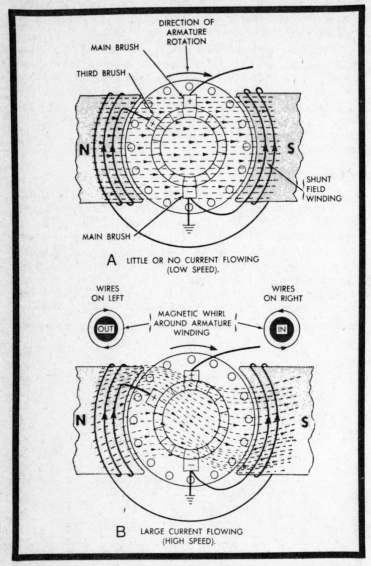

Fig. 9-20. Third-brush generator. A speed and output increase, the field's excitation current drops because of magnetic whirl.

### Third-Brush Generator

The third brush is connected to the field coils as shown in Fig. 9-20. At low rotational speeds the magnetic lines of force bisect the armature in a uniform manner. But as speed and output increase, the field distorts. The armature generates its own field since a magnetic field is created at right angles to a conductor when current flows through it. The resulting field distortion, sometimes called *magnetic whirl*, places the loops feeding the third brush in an area of relative magnetic weakness. Consequently, less current is generated in these loops, and the output to the fields is lessened. The fields become correspondingly weaker, and the total generator output as defined by the negative and positive brushes remains stable or declines.

The output depends upon the position of the third brush, which can be moved in or out of the distorted field to adjust output for anticipated loads. Moving the brush in the direction of armature rotation increases the output; moving it against the direction of rotation reduces the output. When operated independently from the external circuit, the brushes must be grounded to protect the windings. Many of these generators have a fuse in series with the fields.

## DC MOTORS

One of the pecularities of a direct current generator is that it will "motor" if the brushes are connected to a voltage source. In like manner a direct current motor will "gen" if the armature is rotated by some mechanical means. Some manufacturers of small-engine accessories have taken advantage of this phenomenon to combine both functions in a single housing. One such system is the *Dynastart* system employed on many single-cylinder Hatz engines (shown schematically in Fig. 9-21).

Figure 9-22 illustrates the motor effect. In the top sketch the conductor is assumed to carry an electric current toward you. The magnetic flux around it travels in a clockwise direction. Magnetic lines of force above the conductor are distorted and stretched. Since the lines of force have a strong elastic tendency to shorten, they push against the conductor. Placing a loop of wire in the field (bottom sketch), instead of a single conductor, doubles the motor effect. Current goes in the right half of the loop and leaves at the left. The interaction of the fields causes the loop (or, collectively, the armature) to turn counterclockwise. Reversing the direction of the current would cause torque to be developed in a clockwise direction.

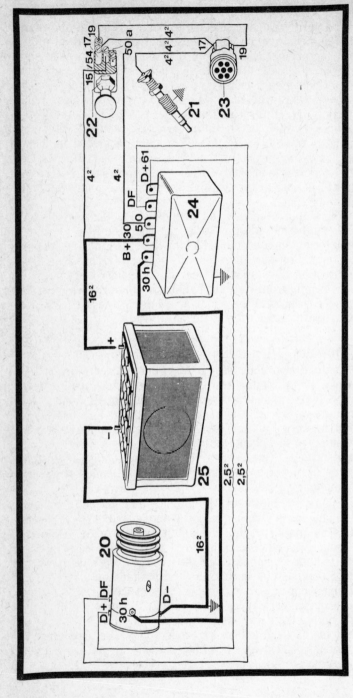

Fig. 9-21. Combination generator and starter motor. (Courtesy Teledyne Wiscousin Motor.)

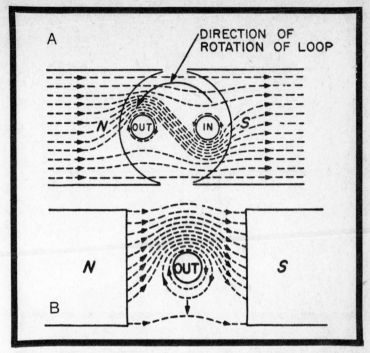

Fig. 9-22. Motor effect created by current bearing conductor in a magnetic field.

Starter motors are normally wired with the field coils in series with the armature (Fig. 9-23). Any additional load added to a series motor will cause more current in the armature and correspondingly more torque. Since this increased current must pass through the series field, there will be a greater flux. Speed changes rapidly with load. When the rotational speed is low, the motor produces its maximum torque. A starter may draw 300A during cranking, more than twice that figure at stall.

## STORAGE BATTERIES

Storage batteries accumulate electrical energy and release it upon demand. The familiar lead–acid battery was invented more than 100 years ago by Gaston Plante. It suffers from poor energy density (watt-hours per pound) and poor power density (watts per pound). The average life is said to be in the neighborhood of 360 complete charge–discharge cycles. During charging, the lead–acid battery shows an efficiency of about 75%. That is, only three-quarters of the input can be retrieved.

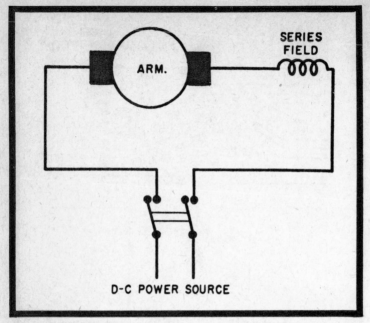

Fig. 9-23. Series field and armature connections typical of starter motors.

Yet it remains the only practical alternative for automotive, marine, and most stationary engine applications. Sodium—sulfur, zinc—air, lithium—halide, and lithium—chlorine batteries all have superior performance, but are impractical by reason of cost and, in some cases, the need for complex support systems.

The lead—acid battery consists of a number of cells; (hence the name *battery*) connected in series. Each fully charged cell is capable of producing 2.2V. The number of cells fixes the output: 12V batteries have six cells 24V batteries have 12. The cells are enclosed in individual compartments in a rubberoid or (currently) high-impact plastic case. The compartments are sealed from each other and, with the exception of Delco and other "zero maintenance" types, open to the atmosphere. The lower walls of the individual compartments extend below the plates to form a sediment trap. Filler plugs are located on the cover and may be combined with wells or other visual indicators to monitor the electrolyte level.

The cells consist of a series of lead plates (Fig. 9-24) connected by internal straps. Until recently the straps were routed over the top of the case, making convenient test points for the technician. Unfortunately, these straps leaked current

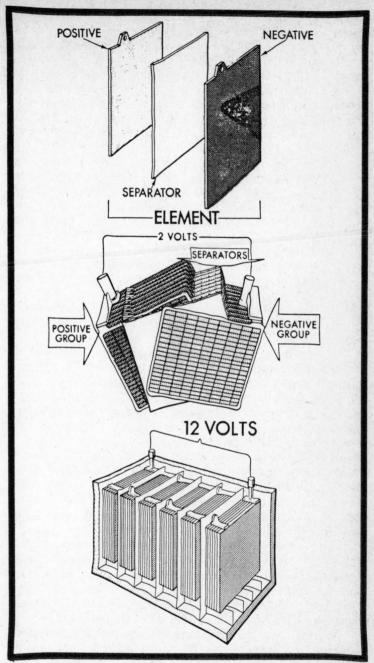

Fig. 9-24. Lead – acid storage battery construction.

and were responsible for the high self-discharge rates of these batteries.

The plates are divided into positive and negative groups and separated by means of plastic or fiberglass sheeting. Some very large batteries, which are built almost entirely by hand, continue to use fir or Port Orford cedar separators. A few batteries intended for vehicular service feature a loosely woven fiberglass padding between the separators and positive plates. The padding gives support to the lead filling and reduces damage caused by vibration and shock.

Both sets of plates are made of lead. The positive plates consist of a lead gridwork which has been filled with lead oxide paste. The grid is stiffened with a trace of antimony. Negative plates are cast in sponge lead. The plates and separators are immersed in a solution of sulfuric acid and distilled water. The standard proportion is 32% acid by weight.

The level of the electrolyte drops in use because of evaporation and hydrogen loss. (Sealed batteries have vapor condensation traps molded into the roof of the cells.) It must be periodically replenished with distilled water. Nearly all storage batteries are shelved dry, and filled upon sale. Once the plates are wetted, electrical energy is stored in the form of chemical bonds.

When a cell is fully charged the negative plates consist of pure sponge lead (Pb in chemical notation), and the positive plates are lead dioxide ($PbO_2$, sometimes called lead peroxide). The electrolyte consists of water ($H_2O$) and sulfuric acid ($H_2SO_4$). The fully charged condition corresponds to drawing A in Fig. 9-25. During discharge both sponge lead and lead dioxide become lead sulfate ($PbSO_4$). The percentage of water in the electrolyte increases since the $SO_4$ radical splits off from the sulfuric acid to combine with the plates. If it were possible to completely discharge a lead–acid battery, the electrolyte would be safe to drink. In practice, batteries cannot be completely discharged in the field. Even those which have lain about junkyards for years still have some charge.

During the charge cycle the reaction reverses. Lead sulfate is transformed back into lead and acid. However, some small quantity of lead sulfate remains in its crystalline form and resists breakdown. After many charge–discharge cycles the residual sulfate permanently reduces the battery's output capability. The battery then is said to be *sulfated*.

The rate of self-discharge is variable and depends upon ambient air temperature, the cleanliness of the outside surfaces of the case, and humidity. In general it averages 1% a day. Batteries which have discharged below 70% of full charge

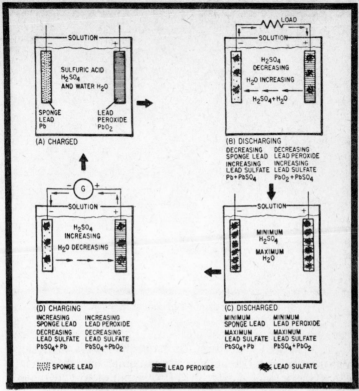

Fig. 9-25. Chemical action in a lead – acid cell.

may sulfate. Sulfation becomes a certainty below the 70% mark. In addition to the possibility of damage to the plates, a low charge brings the freezing point of the electrolyte to near 32°F.

Temperature changes have a dramatic effect upon the power density. Batteries are warm-blooded creatures and perform best at room temperatures. At 10°F the battery has only half of its rated power.

## SWITCHES

Switches neither add energy to the circuit nor take it out. Their function is to give flexibility by making or breaking circuits or by providing alternate current paths. They are classified by the number of *poles* (movable contacts) and *throws* (closed positions). Figure 9-26 illustrates four standard switch types in schematic form.

A single-pole, single-throw switch is like an ordinary light switch in that it makes and breaks a single circuit with one

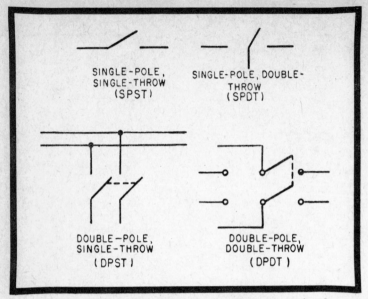

SINGLE-POLE,
SINGLE-THROW
(SPST)

SINGLE-POLE, DOUBLE-
THROW
(SPDT)

DOUBLE-POLE,
SINGLE-THROW
(DPST)

DOUBLE-POLE,
DOUBLE-THROW
(DPDT)

Fig. 9-26. Schematic symbols of commonly used switches.

movement from the rest position. Single-pole, double-throw (spdt) switches have three terminals, but control a single circuit with each throw. As one is completed, the other is opened. You can visualize an spdt as a pair of spst switches in tandem. Double-pole, double-throw switches control two circuits with a single movement. Think of the dpdt as two dpst switches combined, but working so that when one closes the other opens.

### Self-Actuating Switches

Not all switching functions are done by hand. Some are too critical to trust to the operator's alertness. In general two activating methods are used. The first depends upon the effect of heat on a bimetallic strip or disc to open or close contacts. Heat may be generated through electrical resistance or, as is more common, from the engine coolant. The bimetallic element consists of two dissimilar metals bonded back-to-back. Since the coefficient of expansion is different for the metals, the strip or disc will deform inward, in the direction of the least expansive metal. Figure 9-27 illustrates this in a flasher unit. Other switches operate by the pressure of a fluid acting against a diaphragm. The fluid may be air, lube oil, or brake fluid. The most common use of such switches is as oil pressure sensors (Fig. 9-28).

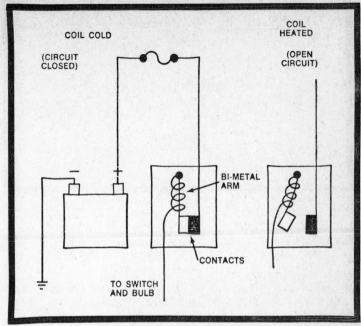

Fig. 9-27. Bimetallic switch operation.

## Relays

Relays are electrically operated switches. They consist of an electromagnet, a movable armature, a return spring, and one or more contact sets. Small currents excite the coil, and the resulting magnetic field draws the armature against the coil. This linear movement closes or opens the contacts. The attractiveness of relays is that small currents—just enough to

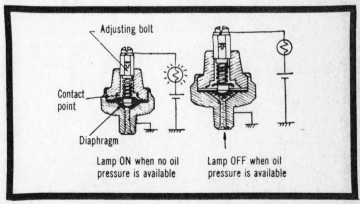

Fig. 9-28. Diaphragm switch (oil pressure).

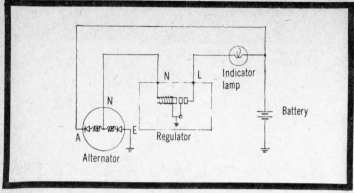

Fig. 9-29. Relay-tripped charging indicator lamp.

excite the electromagnet—can be used to switch large currents. You will find relays used to activate charging indicator lamps (Fig. 9-29), some starter motors, horns, remotely controlled air trips, and the like.

Solenoids are functionally similar to relays. The distinction is that the movable armature has some nonelectrical chore, although it may operate switch contacts in the bargain. Many starter motors employ a solenoid to lever the pinion gear into engagement with the flywheel. We will consider several of these starters in the next chapter.

**Circuit Protective Devices**

Fuses and fusible links are designed to vaporize and open the circuit during overloads. Some circuits are routed through

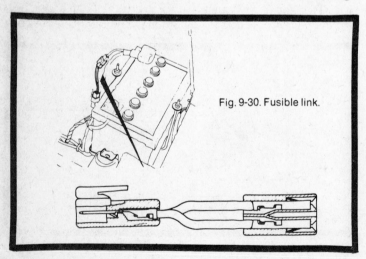

Fig. 9-30. Fusible link.

central fuse panels. Others may be protected by individual fuses spliced into the circuit at strategic points.

*Fusible links* are appearing with more frequency as a protection for the battery—regulator circuit (Fig. 9-30). Burnout is signaled by swollen or discolored insulation over the link. New links may be purchased in bulk for soldered joints, or in precut lengths for use with quick-disconnects. Before replacing the link, locate the cause of failure, which will be a massive short in the protected circuit.

On the other hand, a blown fuse is no cause for alarm. Fuses fatigue and become resistive with age. And even the best regulated charging circuit is subject to voltage and current spikes which may take out a fuse but otherwise are harmless. Chronic failure means a short or a faulty regulator.

### Capacitors (Condenser)

Two centuries ago it was thought that Leyden jars for storing charges actually condensed electrical "fluid." These glass jars, wrapped with foil on both the inner and outer surfaces, became known as *condensers*. The term lingers in the vocabulary of automotive and marine technicans, although *capacitor* is more descriptive of the device's capacity for storing electricity.

A capacitor consists of two plates separated by an insulator or *dielectric*. (Fig. 9-31). In most applications the plates and dielectric are wound upon each other to save space.

A capacitor is a kind of a storage tank for electrons. In function it is similar to an accumulator in hydraulic circuits. Electrons are attracted to the negative plate by the close

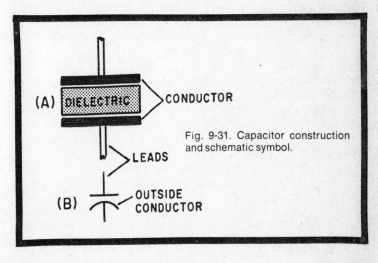

Fig. 9-31. Capacitor construction and schematic symbol.

proximity of the positive plate. Storing the electrons requires energy, which is released when the capacitor is allowed to discharge. Shorting the plates together sends free electrons out of the negative plate and to the positive side. There is a to-and-fro shuttle of electrons between the plates until the number of electrons on the plates is equalized. The number of oscillations depends upon circuit resistance and *reactance* (or the reluctance of the capacitor to charge because of the mutual repulsion of electrons on the negative plate).

Capacitors block direct current, but pass alternating current. The electrons do not physically penetrate the dielectric. The distortions in the orbits of the electrons in the dielectric displace or attract electrons at the plates. The effect on alternating current in a circuit is often as though the dielectric did not exist at least at the higher frequencies.

Capacitors are found at the generator or alternator, where they shunt high-frequency alternating current to ground to muffle radio interference.

## SOLID-STATE COMPONENTS

Although the transistor was invented in the Bell Laboratories in 1947, it and its progeny remain a mystery to most mechanics. These components are used in all alternators to rectify alternating current to direct current and may be found in current regulators produced by CAV, Lucas, Bosch, and Delco. In the future they will have a much wider use and may substitute for relays.

Solid-state components are made of materials called semiconductors, which fall midway in the resistance spectrum between conductors and insulators. In general these components remain nonconductive until subjected to a sufficient voltage.

Silicon and germanium are the most used materials for semiconductors. Semiconductor crystals for electronics are artificial crystals, grown under laboratory conditions. In the pure state these crystals are electrically just mediocre conductors. But if minute quantities of impurities are added in the growth state, the crystals acquire special electrical characteristics that make them useful for electronic devices. The process of adding impurities is known as *doping*.

When silicon or germanium is doped with arsenic or phosphorus, the physical structure remains crystalline. But for each atom of impurity which combines with the intrinsic material, one free electron becomes available. This material has electrons to carry current and is known as $n$-type semiconductor material.

On the other hand the growing crystals may be doped with indium, aluminum, boron, or certain other materials and result in a shortage of electrons in the crystal structure. "Missing" electrons can be thought of as electrical *holes* in the structure of the crystal. These holes accept electrons which blunder by, but the related atom quickly releases the newcomer. When a voltage is applied to the semiconductor, the holes appear to drift through the structure as they are alternately filled and emptied. Crystals with this characteristic are known as p-type materials, since the holes are positive charge carriers.

## Diode Operation

When p-type material is mated with n-type material, we have a *diode*. This device has the unique ability to pass current in one direction and block it in the other. When the diode is not under electrical stress, p-type holes in the p-material and electrons in the n-material complement each other at the interface of the two materials. This interface, or *junction*, is electrically neutral, and presents a barrier to the charge carriers. When we apply negative voltage to the n-type material, electrons are forced out of it. At the same time, holes in the p-type material migrate toward the direction of current flow. This convergence of charge carriers results in a steady current in the circuit connected to the diode. Applying a negative voltage to the n-type terminal so the diode will conduct is called forward biasing.

If we reverse the polarity—that is, if we apply positive voltage to the n-type material—electrons are attracted out of it at the same time, holes move away from the junction on the p side. Since charge carriers are thus made to avoid the junction, no current will pass. The diode will behave as an insulator until the reverse-bias voltage reaches a high enough level to destroy the crystal structure.

Diodes are used to convert alternating current from the generator to pulsating direct current. An alternator typically has three pairs of diodes mounted in heatsinks (heat absorbers).

Diodes do have some resistance to forward current and so must be protected from heat by means of shields and sinks. In a typical alternator the three pairs of diodes which convert alternating current to pulsating direct current are pressed into the aluminum frame. Heat generated in operation passes out to the whole generator.

## Solid-State Characteristics

As useful as solid-state components are, they nevertheless are subject to certain limitations. The failure mode is absolute. The device either works or it doesn't. And in the event of failure, no amount of circuit juggling or tinkering will restore operation. The component must be replaced. Failure may be spontaneous—the result of manufacturing error compounded by the harsh environment under the hood—or it may be the result of faulty service procedures. Spontaneous failure usually occurs during the warranty period. Service-caused failures tend to increase with time and mileage as the opportunities for human error increase.

These factors are lethal to diodes and transistors:

- High inverse voltage resulting from wrong polarity: Jumper cables connected backwards or a battery installed wrong will scramble the crystalline structure of these components.
- Vibration and mechanical shock: Diodes must be installed in their heatsinks with the proper tools. A $2 \times 4$ block is not adequate.
- Short circuits: Solid-state devices produce heat in normal operation. A transistor, for example, causes a 0.3V—0.7V drop across the collector−emitter terminals. This drop, multiplied by the current, is the wattage consumed. Excessive current will overheat the device.
- Soldering: The standard practice is for alternator diodes to be soldered to the stator leads. Too much heat at the connection can destroy the diodes.

# Starting and Generating Systems

**10**

The diesel engines we are concerned with are almost invariably fitted with electric starter motors. A number of engines, used to power construction machinery and other vehicles which are expected to stand idle in the weather, employ a gasoline engine rather than an electric motor as a starter. A motor demands a battery of generous capacity and a generator to match.

Starting a cold engine can be somewhat frustrating, particularly if the engine is small. The surface/volume ratio of the combustion space increases disproportionately as engine capacity is reduced. The heat generated by compression tends to dissipate through the cylinder and head metal. In addition, cold clearances may be such that much of the compressed air escapes past the piston rings. Other difficulties include the effect of cold on lube and fuel oil viscosity. The spray pattern coarsens, and the drag of heavy oil between the moving parts increases.

Starting has more or less distinct phases. Initial or *breakaway* torque requirements are high since the rotating parts have settled to the bottom of their journals and are only marginally lubricated. The next phase occurs during the first few revolutions of the crankshaft. Depending upon ambient temperature, piston clearances, lube oil stability, and the like, the first few revolutions of the crankshaft are free of heavy compressive loads. But cold oil is being pumped to the journals which collects and wedges between the bearings and the shafts. As the shafts continue to rotate, the oil is heated by friction and thins, progressively reducing drag. At the same time, cranking speed increases and compressive loads become significant. The engine accelerates to *firing speed*. The

Fig. 10-1. If the ignition paper is clamped properly it will support the weight of the holder. (Courtesy Wisconsin Teledyne Motor.)

duration from breakaway to firing speed depends upon the capacity of the starter and battery, the mechanical condition of the engine, lube oil viscosity, ambient air temperature, the inertia of the flywheel, and the number of cylinders. A single-cylinder engine is at something of a disadvantage since it cannot benefit from the expansion of other cylinders. Torque demands are characterized by sharp peaks.

## STARTING AIDS

It is customary to include a cold-starting position at the rack. This position provides extra fuel to the nozzles and makes combustion correspondingly more likely.

Other starting aids involve heat. One of the most primitive and effective is to introduce chemically treated papers (punks) into the cylinders. Some care must be exercised to be certain that the paper is held firmly by the holder (Fig. 10-1). Should it break loose and lodge under a valve, teardown may be required. In some cases you may be able to blow the paper free with air.

Lube oil and water immersion heaters are available which can be mounted permanently on the engine. Lube oil heaters are preferred and may be purchased from most engine builders. Good results may be had by heating the oil from an external heater mounted below the sump. Use an approved type to minimize the fire hazard. Alternatively, one may drain the oil upon shutdown and heat it before starting. The same may be done with the coolant, although temperatures in both cases should be kept well below the boiling temperature of water to prevent distortion and possible thermal cracking.

Another method, favored by Mercedes-Benz, Hatz, Chrysler-Nissan, and many other builders, is to employ *glow plugs* (Fig. 10-2). These plugs are threaded into the individual cylinders and are heated by battery current. Operating temperatures of 1500°F are reached in seconds. The plugs are wired in parallel and have a current draw of 5−6A. To inspect, remove the plug from the cylinder and connect to a circuit main. The plug should glow red hot. Resistance varies with make and model, but it should be very low. A reading of 1.5 ohms is normal.

Starting fluid may be used in the absence of intake air heaters. In the old days a mechanic poured a spoonful of ether on a burlap rag and placed it over the air intake. This method is not the safest nor the most consistent: Too little fluid will not start the engine, and too much can cause severe detonation or an intake header explosion. Aerosol cans are available for injection directly into the air intake. Use as directed in a well ventilated place.

More sophisticated methods include pumps and metering valves in conjunction with pressurized containers of starting fluid. Figure 10-3 illustrates a typical metering valve. The valve is tripped only once during each starting attempt, to forestall explosion. Caterpillar engines are sometimes fitted with a one-shot starting device consisting of a holder and needle. A capsule of fluid is inserted in the device and the needle pierces it, releasing the fluid.

The starter motor should not be operated for more than a few seconds at a time. Manufacturers have different recommendations on the duration of cranking, but none suggests that the starter button be depressed for more than 30 seconds. Allow a minute or more between bouts for cooling and battery recovery.

If extensive cold weather operation is intended or if the engine will be stopped and started frequently, it is wise to add one or more additional batteries wired in parallel. Negative-to-negative and positive-to-positive connections do

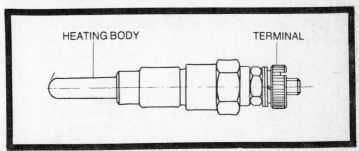

Fig. 10-2. Sheathed-type glow plug. (Courtesy Marine Engine Div., Chrysler Corp.)

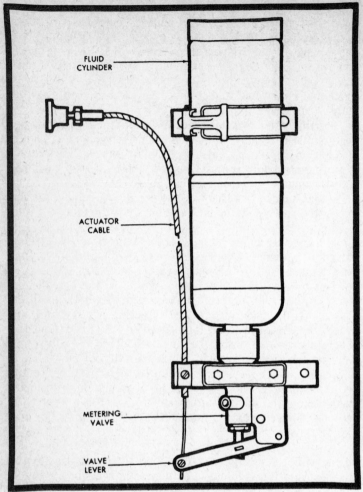

FLUID
CYLINDER

ACTUATOR
CABLE

METERING
VALVE

VALVE
LEVER

Fig. 10-3. Quick-start unit. (Courtesy Detroit Diesel Allison.)

not alter the output voltage, but add the individual battery capacities.

## WIRING

In Chapter 9, Fig. 9-21 showed a wiring layout which incorporated a *Dynastart* combination starter motor and generator. Figure 10-4 shows a more typical example; the starter motor and generator are distinct units. The next drawing illustrates a circuit with an alternator instead of a direct current generator. All of these circuits are similar in configuration. Power from the battery goes to the starter

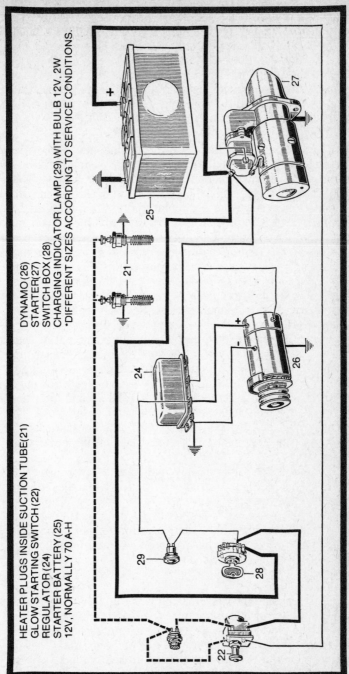

HEATER PLUGS INSIDE SUCTION TUBE (21)
GLOW STARTING SWITCH (22)
REGULATOR (24)
STARTER BATTERY (25)
12V, NORMALLY 70 A-H

DYNAMO (26)
STARTER (27)
SWITCH BOX (28)
CHARGING INDICATOR LAMP (29) WITH BULB 12V, 2W
*DIFFERENT SIZES ACCORDING TO SERVICE CONDITIONS.

Fig. 10-4. Circuit with starter motor and direct current generator. (Courtesy Teledyne Wisconsin Motor.)

227

through a relay which is triggered by the starter switch. The Ford layout in Fig. 10-5 employs a relay and a solenoid mounted atop the starter motor frame. The solenoid also carries contacts. Power to the battery leaves the generator, goes to the regulator (which is internal in the Ford design), through an optional ammeter or lamp, and to the battery from a tie point at the starter solenoid. One or more fuses and a fusible link may be included in the circuit.

These diagrams are not to scale, and the routing of the wires has been simplified. If your engine has been wired professionally, most of the conductor will be gathered in a harness. An exact depiction of engine wiring would not be helpful to the technician, who needs to know where the wires terminate, and not what particular routes they may take in getting there. Some manufacturers simplify their diagrams further, as in the example shown in Fig. 10-6.

In addition to the pattern of wiring, a diagram may indicate the color codes used for particular circuits and the wire size. Color coding has not been standardized, although there is general agreement that black should be used for ground wires. Most manufacturers stay with primary colors and avoid the exotic, which would not only confuse mechanics, but make translation into other languages problematic. Thus, if a wire is shown as *O*, you can be sure the color is orange, and not ochre or opal. *Black* is often abbreviated *BK* to distinguish it from *brown* or *blue*. Multicolored wires are used for many circuits. The first color is the base and the second the trace. Thus *YG* means a yellow conductor with a green trace. The wiring may be divided functionally into circuit blocks. Starting, charging, instrumentation, and lighting are typical divisions.

Sometimes each circuit block will have its own base color. As the circuit develops, the color of the trace changes in some predetermined sequence. For example, lighting circuits for Nissan engines are base-colored red. As the circuit progresses the trace becomes white, black, yellow, green, and blue. Instrument circuits are base-colored yellow and follow the same progression of trace colors.

## Wiring Repairs

Most repairs involve cleaning and tightening terminals and taping abraded insulation. Cotton friction tape has given way to the newer vinyl type, which has a claimed dielectric strength of 600V per layer. In marine and other severe environments, it is a wise precaution to seal the edges of the tape with a thin coating of silicone cement.

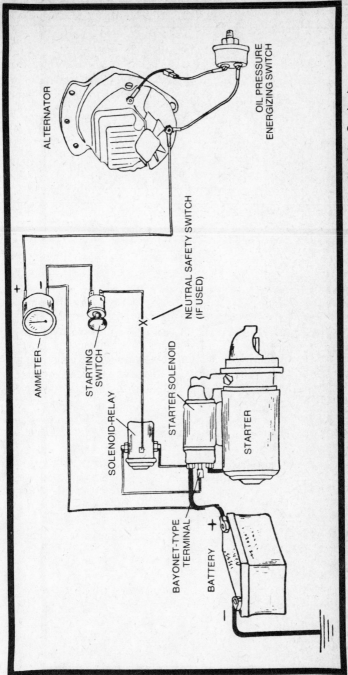

Fig. 10-5. Circuit with starter motor and alternator. (Courtesy Lehman Manufacturing Co., Inc.)

ALTERNATOR

OIL PRESSURE
ENERGIZING SWITCH

NEUTRAL SAFETY SWITCH
(IF USED)

AMMETER

STARTING
SWITCH

SOLENOID-RELAY

STARTER SOLENOID

STARTER

BAYONET-TYPE
TERMINAL

BATTERY

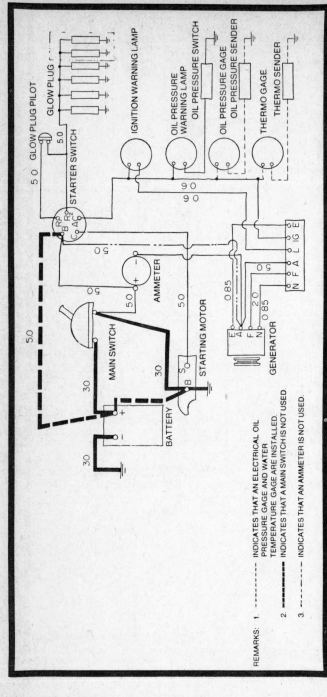

Fig. 10-6. Wiring diagram for Chrysler Nissan SD22 and SD33 engines. The values given on the wiring diagram are reference values of the sectional areas of the wiring used. (Courtesy Marine Engine Div., Corp.)

REMARKS: 1. — — — INDICATES THAT AN ELECTRICAL OIL PRESSURE GAGE AND WATER TEMPERATURE GAGE ARE INSTALLED.

2. ▬▬▬ INDICATES THAT A MAIN SWITCH IS NOT USED.

3. — — — INDICATES THAT AN AMMETER IS NOT USED.

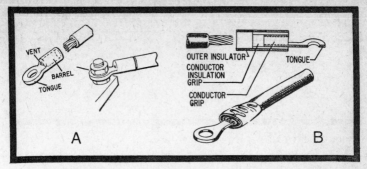

Fig. 10-7. (A) Solder-type terminal lug. (B) Crimp-on terminal lug.

All connections should be made with terminal lugs. Tongue and barrel size vary with the wire gage, and terminals should be purchased accordingly. Solder-type lugs (Fig. 10-7) can make mechanically strong, almost zero-resistance joints, but in recent years these have been replaced by solderless lugs in production and field repairs. Solderless lugs are expensive and require a special crimping tool to install, but the results are uniform.

Splices should be made with wires twisted as shown in Fig. 10-8. A variant which is becoming popular in the automotive trades is illustrated in Fig. 10-9. The insulation is cut back ¾ in. or so, and the strands are splayed open. The strands are interleaved and given a twist. Then solder is applied to the joint.

The cross-sectional area of the conductor (not the insulation) is important because it determines the current-carrying capacity of the wire. In the mechanical

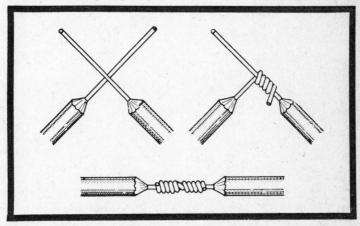

Fig. 10-8. Western Union splice.

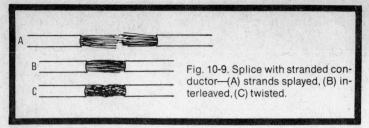

Fig. 10-9. Splice with stranded conductor—(A) strands splayed, (B) interleaved, (C) twisted.

trades the most commonly used measure of wire size is the SAE (Society of Automotive Engineers) gage. Like many measures, *f*-stop numbers, shotgun bores, etc, the larger sizes are assigned a lower number. Thus, a 10-gage conductor is larger than a 12-gage. The metric system is more straightforward—assigned numbers correspond to the approximate cross-sectional area in square millimeters. Figure 10-10 compares these two systems and gives the current-carrying capacity of the various sizes.

As indicated in the table, the standard wiring is stranded copper with vinyl insulation. This wire is known as *primary wire* or automotive hookup wire. The stranded construction gives it flexibility and resistance to weakening caused by flexing. Polyvinylchloride insulation is good to 105°C. In

| SAE wire size | Sectional area | | Outer dia. | | Conductor resistance ($\Omega$/km) | Allowable continuous load current (A) |
|---|---|---|---|---|---|---|
| | Circular mils | mm$^2$ | in. | mm | | |
| 20 | 1,094 | 0.5629 | 0.040 | 1.0 | 32.5 | 7 |
| 18 | 1,568 | 0.8846 | 0.050 | 1.2 | 20.5 | 9 |
| 16 | 2,340 | 1.287 | 0.060 | 1.5 | 14.1 | 12 |
| 14 | 3,777 | 2.091 | 0.075 | 1.9 | 8.67 | 16 |
| 12 | 5,947 | 3.297 | 0.090 | 2.4 | 5.50 | 22 |
| 10 | 9,443 | 5.228 | 0.115 | 3.0 | 3.47 | 29 |
| 8 | 15,105 | 7.952 | 0.160 | 3.7 | 2.28 | 39 |
| 6 | 24,353 | 13.36 | 0.210 | 4.8 | 1.36 | 88 |
| 4 | 38,430 | 20.61 | 0.275 | 6.0 | 0.871 | 115 |
| 2 | 63,119 | 35.19 | 0.335 | 8.0 | 0.510 | 160 |
| 1 | 80,010 | 42.73 | 0.375 | 8.6 | 0.420 | 180 |
| 0 | 100,965 | 54.29 | 0.420 | 9.8 | 0.331 | 210 |
| 2/0 | 126,822 | 63.84 | 0.475 | 10.4 | 0.281 | 230 |
| 3/0 | 163,170 | 84.96 | 0.535 | 12.0 | 0.211 | 280 |
| 4/0 | 207,740 | 109.1 | 0.595 | 13.6 | 0.165 | 340 |

Fig. 10-10. Specifications for vinyl-insulated automotive wire. (Courtesy Marine Engine Div., Chrysler Corp.)

high-temperature applications you can use Teflon, which can withstand twice the heat.

## Soldering

Soldering is an art which requires some practice to master. Begin with a soldering iron or gun of about 250W rating. This wattage rating will prove more than adequate for most jobs. However, such an iron will not be large enough to solder battery terminals to their cables; for this you will need a torch. By the same token, a 250W iron must be used with great care when soldering diode leads. The heat generated can easily ruin the diode. For small jobs a rechargeable battery-powered iron such as that shown in Fig. 10-11 is handy.

The tip of a nonplated soldering iron should be dressed with a file down to virgin copper and *tinned*, or coated with molten solder. Silver solder is preferable for tinning since it melts at a higher temperature than the lead type and so protects the tip from corrosion. The tip should be periodically retightened since heating and cooling cycles tend to loosen it.

Solder used for electrical work is available in these grades: 40/60, 50/50, and 60/40. The first figure is the percent of tin, and the other is the percent of lead. In general, employ 60/40 solder. The melting temperature is lower than for the other grades, which means less chance of damage to insulation

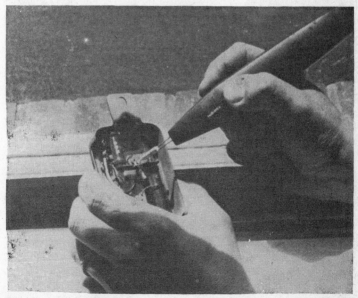

Fig. 10-11. A Wahl battery-powered soldering iron. With the proper tip this iron will solder three No. 10 wires.

and delicate components from thermal spillover; and the solder hardens more quickly. The $^1/_{16}$ in. diameter size is most convenient.

Soldering produces a molecular bond between materials to be joined. All oxides must be removed and the connections scraped bright. A flux is required to prevent oxidation during the soldering process. Acid flux is used for radiator, fuel tank, and other repairs where mechanical strength is a factor. It cannot be used on current-carrying joints, since the flux creates resistance. Specify rosin flux for electrical work. Generally the rosin is in the core of the solder, so that a separate flux is not necessary.

The following rules have been developed from experience and out of a series of experiments conducted by the military.

- Use a minimum amount of solder.
- Wrapping terminal lugs and splice ends with multiple turns of wire does not add to the mechanical strength of the joint and increases the heat requirement. Wrap only to hold the joint while soldering.
- Heat the connection, not the solder. When the parts to be joined are hot enough solder will flow into the joint.
- Do not move the parts until the solder has hardened. Movement while the solder is still plastic will produce a highly resistive "cold" joint.
- Use only enough heat to melt the solder. Excessive heat can damage nearby components and can crystallize the solder.
- Allow the joint to air-cool. Dousing a joint with a water to cool it weakens the bond.

## STARTER CIRCUITS

Before assuming that the motor is at fault, check the battery and cables. The temperature-corrected hydrometer reading should be at least 1.240, and no cell should vary from the average of the others by more than 0.05 point. See that the battery terminal connections are tight and free of corrosion.

Excessive or chronic starter failure may point to a problem that is outside the starter itself. It could be caused by an engine that is out of tune and which consequently requires long cranking intervals.

### Starter Circuit Tests

There are several methods which you can use to check the starting-circuit resistance. One method is to open all the connections, scrape bright, and retighten. Another method requires a low-reading ohmmeter of the type sold by Sun

Electric and other suppliers for the automotive trades. But most mechanics prefer to test by voltage drop.

Connect a voltmeter as shown in Fig. 10-12. The meter shunts the positive, or *hot* battery post and the starter motor. With the meter set on a scale above battery voltage, crank. Full battery voltage means an open in the circuit.

If the starter functions at all, the reading will be only a fraction of this. Expand the scale accordingly. A perfect circuit will give a zero voltage drop since all current goes to the battery. In practice some small reading will be obtained. The exact figure depends upon the current draw of the starter and varies between engine and starter motor types. As a general rule, subject to modification by experience, a 0.5V drop is normal. Much more than this means: (1) resistance in the cable, (2) resistance in the connections (you can localize this by repeating the test at each connection point), or (3) resistance in the solenoid.

Figure 10-13 shows the connections for the ground-side check. A poor ground, and consequent high voltage on the meter, can occur at the terminals, the cable, or between the starter motor and engine block. If the latter is the case, remove the motor and clean any grease or paint from the mounting flange.

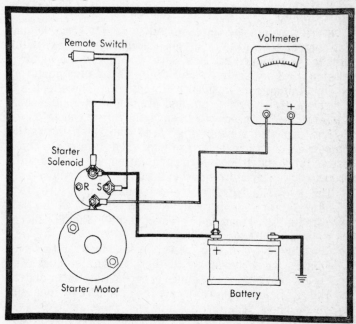

Fig. 10-12. "Hot" side voltage drop test.

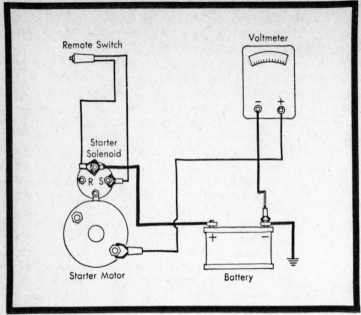

Fig. 10-13. Ground circuit test.

## Starter Motors

Starter motors are series-wound; i.e., they are wound so that current enters the field coils and goes to the armature through the insulated brushes. Since a series-wound motor is characterized by high no-load rpm, some manufacturers employ limiting coils in shunt with the fields. The effect is to govern the free-running rpm and prolong starter life should the starter be energized without engaging the flywheel.

The exploded view in Fig. 10-14 illustrates the major components of a typical starter motor. The frame (No. *1*) has several functions. It locates the armature and fields, absorbs torque reaction, and forms part of the magnetic circuit.

The field coils (No. *2*) are mounted on the pole shoes (No. *3*) and generate a magnetic field which reacts with the field generated in the armature to produce torque. The pole pieces are secured to the frame by screws.

The armature (No. *4*) consists of a steel form and a series of windings which terminate at the commutator bars. The shaft is integral with it and splined to accept the starter clutch.

The end plates (*5* and *6*) locate the armature by means of bronze bushings. The commutator end plate doubles as a mounting fixture for the brushes, while the power takeoff side segregates the starter motor from the clutch.

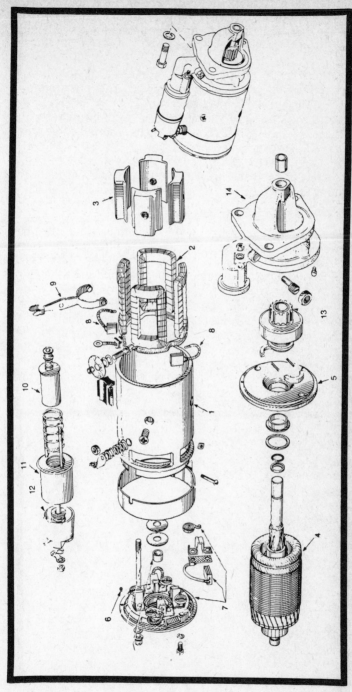

Fig. 10-14. Starter in exploded view and as assembled. (Courtesy Lehman Manufacturing Co., Inc.)

The insulated (hot) brushes (No. 7) provide a current path from the field coils through the commutator and armature windings to the grounded brushes (No. 8).

Engagement of this particular starter is by means of a yoke (No. 9) which is pivoted by the solenoid plunger (No. 10) in response to current flowing through the solenoid windings (No. 11). Movement of the plunger also trips a relay (No. 12) and energizes the motor. The pinion gear (No. 13) meshes with the ring gear on the rim of the flywheel. The pinion gear is integral with an overrunning clutch.

The starter drive housing supports the power takeoff end of the shaft and provides an accurately machined surface for mounting the starter motor to the engine block or bell housing.

## Brushes

Before any serious work can be done, the starter must be removed from the engine, degreased, and placed on a clean bench. Disconnect one or both cables at the batter to prevent sparking; disconnect the cable to the solenoid and the other leads which may be present (noting their position for assembly later); and remove the starter from the flywheel housing. Starters are mounted with a pair of capscrews or studs.

Remove the brush cover, observing the position of the screw or snap, since wrong assembly may short the main cable or solenoid wire (Fig. 10-15). Hitachi starters do not have an inspection band as such. The end plate must be removed for access to the brushes and commutator.

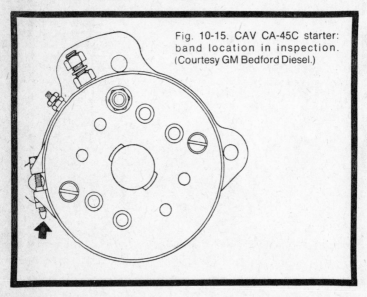

Fig. 10-15. CAV CA-45C starter: band location in inspection. (Courtesy GM Bedford Diesel.)

Brushes are sacrificial items and should be replaced when worn to half their original length. The rate of wear should be calculated so that the wear limit will not be reached between inspection periods. Clean the brush holders and commutator with a preparation such as No. 2002 Freon degreaser. If old brushes are used, lightly file the flanks at the contact points with the holders to help prevent sticking. New brushes are contoured to match the commutator, but should be fitted by hand. Wrap a length of sandpaper around the commutator—do not use emery cloth—and turn in the normal direction of rotation. Remove the paper and blow out the dust.

Try to move the holders by hand. Most are riveted to the end plate and can become loose, upsetting the brush—commutator relationship. With an ohmmeter, test the insulation on the hot-side brush holders (Fig. 10-16). There should be no continuity between the insulated brush holders and the end plate. Brush spring tension is an important and often-overlooked factor in starter performance. To measure it, you will need an accurate gage such as one supplied by Sun Electric. Specifications vary between makes and models, but the spring tension measured at the free (brush) end of the spring should be at least 1½ lb. Some specifications call for 4 lb.

The commutator bars should be examined for arcing, scores, and obvious eccentricity. Some discoloration is normal. If more serious faults are not apparent, buff the bars with a strip of 000 sandpaper.

## Armature

Further disassembly requires that the armature shaft be withdrawn from the clutch mechanism. Some starters employ a snapring at the power takeoff end of the shaft to define the

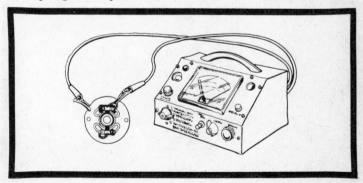

Fig. 10-16. Testing brush holder insulation. (Courtesy Tecumseh Products Co.)

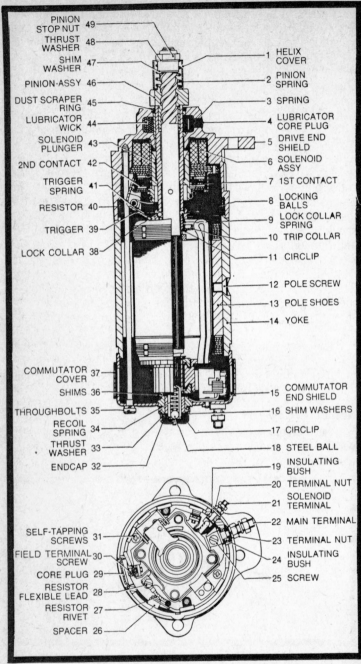

|  | |  |
|---|---|---|
| PINION STOP NUT | 49 | 1 HELIX COVER |
| THRUST WASHER | 48 | 2 PINION SPRING |
| SHIM WASHER | 47 | 3 SPRING |
| PINION-ASSY | 46 | 4 LUBRICATOR CORE PLUG |
| DUST SCRAPER RING | 45 | 5 DRIVE END SHIELD |
| LUBRICATOR WICK | 44 | 6 SOLENOID ASSY |
| SOLENOID PLUNGER | 43 | 7 1ST CONTACT |
| 2ND CONTACT | 42 | 8 LOCKING BALLS |
| TRIGGER SPRING | 41 | 9 LOCK COLLAR SPRING |
| RESISTOR | 40 | 10 TRIP COLLAR |
| TRIGGER | 39 | 11 CIRCLIP |
| LOCK COLLAR | 38 | 12 POLE SCREW |
| | | 13 POLE SHOES |
| | | 14 YOKE |
| COMMUTATOR COVER | 37 | 15 COMMUTATOR END SHIELD |
| SHIMS | 36 | 16 SHIM WASHERS |
| THROUGHBOLTS | 35 | 17 CIRCLIP |
| RECOIL SPRING | 34 | 18 STEEL BALL |
| THRUST WASHER | 33 | 19 INSULATING BUSH |
| ENDCAP | 32 | 20 TERMINAL NUT |
| | | 21 SOLENOID TERMINAL |
| | | 22 MAIN TERMINAL |
| SELF-TAPPING SCREWS | 31 | 23 TERMINAL NUT |
| FIELD TERMINAL SCREW | 30 | 24 INSULATING BUSH |
| CORE PLUG | 29 | 25 SCREW |
| RESISTOR FLEXIBLE LEAD | 28 | |
| RESISTOR RIVET | 27 | |
| SPACER | 26 | |

Fig. 10-17. Typical CAV starter. (Courtesy GM Bedford Diesel.)

outer limit of pinion movement. Others use a stopnut (Fig. 10-17, No. *49*). The majority of armatures may be withdrawn with the pinion and overrunning clutch in place. The disengagement point is at the yoke (Fig. 10-14, No. *9*) and sleeve on the clutch body. Remove the screws holding the solenoid housing to the frame and withdraw the yoke pivot pin. The pin may have a threaded fastener with an eccentric journal, as in the case of Ford designs. The eccentric allows for wear compensation. Or it may be a simple cylinder, secured by a flanged head on one side and a cotter pin or snapring on the other. When the pin is removed there will be enough slack in the mechanism to disengage the yoke from the clutch sleeve, and the armature can be withdrawn. Observe the position of shims—usually located between the commutator and end plate—and, on CAV starters, the spring-loaded ball. This mechanism is shown in Fig. 10-17 as *18* and *34*. It allows a degree of end float so that the armature can recoil if the pinion and flywheel ring gear do not mesh on initial contact.

The armature should be placed in a jig and checked for trueness since a bent armature will cause erratic operation and may, in the course of long use, destroy the flywheel ring gear. Figure 10-18 illustrates an armature chucked between lathe centers. The allowable deflection at the center bearing is 0.002 in., or 0.004 in. on the gage. With the proper fixtures and skill with an arbor press, an armature shaft may be straightened, although it will not be as strong as it was originally and will be prone to bend again. The wiser course is to purchase a new armature.

Make the same check on the commutator. Allowable out-of-roundness is 0.012—0.016 in., or less, depending upon the

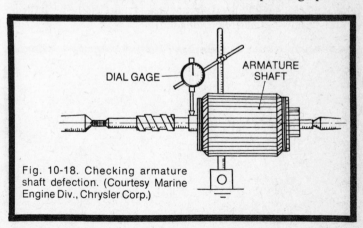

DIAL GAGE

ARMATURE SHAFT

Fig. 10-18. Checking armature shaft defection. (Courtesy Marine Engine Div., Chrysler Corp.)

rate of wear and the intervals between inspection periods. Commutators which have lost their trueness or which have become pitted should be turned on a lathe. The cutting tool must be racked more for copper than for steel. Do not allow the copper to smear into the slots between the segments. Chamfer the end of the commutator slightly. Small imperfections can be removed by chucking the commutator in a drill press and turning against a single-cut file.

It is necessary that the insulation (called, somewhat anachronistically, "mica") be buried below the segment edges; otherwise, the brushes will come into contact with the insulation as the copper segments wear. Undercutting should be limited to 0.015 in. or so. Tools are available for this purpose, but an acceptable job can be done with a hacksaw blade (flattened to fit the groove) and a triangular file for the final bevel cut (Fig. 10-19).

Inspect the armature for evidence of overheating. Extended cranking periods, dragging bearings, chronically low battery charge, or an under-capacity starter will cause the solder to melt at the commutator—armature connections. Solder will be splattered over the inside of the frame. Repairs may be possible since these connections are accessible. Continued overheating will cause the insulation to flake and powder. Discoloration is normal, but the insulation should remain resilient.

Armature insulation separates the nonferrous parts (notably the commutator segments) from the ferrous parts (the laminated iron segments extending to the outer diameter of the armature and shaft). Make three tests with the aid of a 110V continuity lamp or a *megger* (meter for very high resistances). An ordinary ohmmeter is useless to measure the high resistances involved. In no case should the lamp light up or resistance be less than 1 megohm.

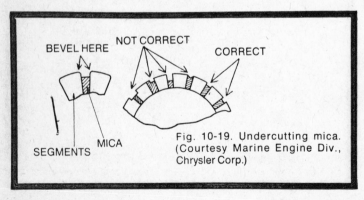

Fig. 10-19. Undercutting mica. (Courtesy Marine Engine Div., Chrysler Corp.)

1. Test between adjacent commutator segments.
2. Test between individual commutator segments and the armature form (Fig. 10-20).
3. Test between armature or commutator segments and the shaft.

It is possible for the armature windings to short, thus robbing starter torque. Place the armature on a growler and rotate it slowly while holding a hacksaw blade over it as shown in Fig. 10-21. The blade will be strongly attracted to the armature segments because of the magnetic field introduced in the windings by the growler. But if the blade does a Mexican hat dance over a segment, you can be sure that the associated winding is shorted.

### Field Coils

After armatures, the next most likely source of trouble is the field coils. Field hookups vary. The majority are connected in a simple series circuit, although you will encounter starters with *split fields*, each pair feeding off its own insulated brush. Field resistance values are not as a rule supplied in shop manuals, although a persistent mechanic can obtain this and

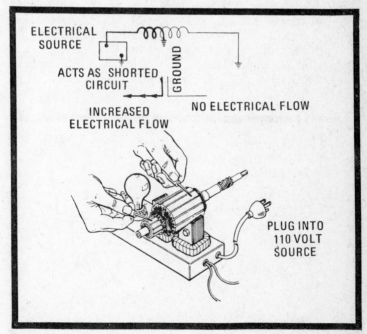

Fig. 10-20. Continuity check between commutator bars and armature segments.

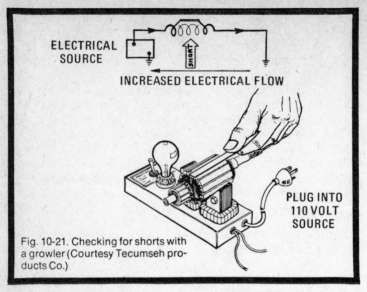

Fig. 10-21. Checking for shorts with a growler (Courtesy Tecumseh products Co.)

other valuable test data by writing the starter manufacturer. Be sure to include the starter model and serial number.

In the absence of a resistance test, which would detect intracoil shorts, the only tests possible are to check field continuity (an ordinary ohmmeter will do) and to check for shorts between the windings and frame. Connect a lamp or megger to the fields and touch the other probe to the frame, as illustrated in Fig. 10-22. Individual fields can be isolated by snipping their leads.

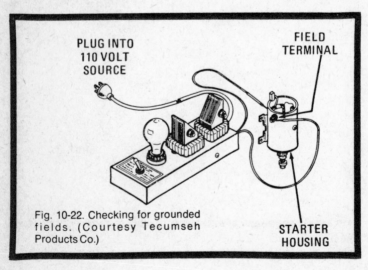

Fig. 10-22. Checking for grounded fields. (Courtesy Tecumseh Products Co.)

The fields are supported by the pole shoes, which, in turn, are secured to the frame by screws. The screws more often than not will be found to have rusted to the frame. A bit of persuasion will be needed, in the form of penetrating oil and hock. Support the frame on a bench fixture and, with a heavy hammer, strike the screwdriver exactly as if you were driving a spike. If this does not work, remove the screw with a cape chisel. However, you may, depending upon the starter make and your supply of junk parts, have difficulty in matching the screw thread and head fillet.

Coat the screw heads with Loctite before installation and torque securely. Be sure that the pole shoe and coil clears the armature and that the leads are tucked out of the way.

## Bearings

The great majority of starters employ sintered bronze bushings. In time these bushings wear and must be replaced to insure proper teeth mesh at the flywheel and prevent armature drag. In extreme cases the bushings may wear down to their bosses so that the shaft rides on the aluminum or steel end plates. The old bushings are pressed out and new ones pressed in. Tools are available to make this job easier, particularly at the blind boss on the commutator end plate.

Without these tools, the bushing can be removed by carefully ridging and collapsing it inward, or by means of hydraulic pressure. Obtain a rod which matches shaft diameter. Pack the bushing with grease and hammer the rod into the boss. Since the grease cannot easily escape between the rod and bushing, it will lift the bushing up and out. Because of the blind boss, new bushings are not reamed.

## Final Tests

The tests described thus far have been *static* tests. If a starter fails a static test it will not perform properly, but passing does not guarantee the starter is faultless. The only sure way to test a starter—or, for that matter, any electrical machine—is to measure its performance under known conditions, and against the manufacturer's specifications or a known-good motor.

No-load performance is checked by mounting the starter in a vise and monitoring voltage, current, and rpm. Figure 10-23 shows the layout. The voltage is the reference for the test and is held to 12V. Current drain and rpm are of course variable, depending upon the resistance of the windings, their configuration, and whether or not speed-limiting coils are provided. You may expect speeds of 4000—7000 rpm and draws

245

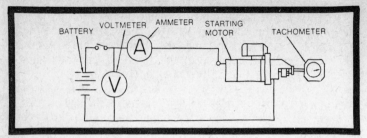

Fig. 10-23. No-load starter test. (Courtesy Marine Engine Div., Chrysler Corp.)

of 60—100A. In this test we are looking for low rpm and excessive current consumption.

The locked-rotor and stall test requires a scale to accept the pinion gear (Fig. 10-24). It must be made quickly, before the insulation melts. Mount the motor securely and lock the pinion. Typically the voltage will drop to half the normal value. Draw may approach or, with the larger starters, exceed 100A.

**SOLENOIDS**

Almost all diesel starters are engaged by means of a solenoid mounted on the frame. At the same time the solenoid plunger moves the pinion, it closes a pair of contacts to complete the circuit to the starter motor. In other words the component consists of a solenoid or *linear motor* and a relay (Fig. 10-25). Some circuits feature a second remotely mounted relay, as shown in Figs. 10-25 and 10-26. The circuit depicted in the earlier illustration was designed for marine use.

The starter switch may be 30 ft or more from the engine. To cut wiring losses a second relay is installed which energizes the piggyback solenoid. The additional relay (*2ST* in Fig. 10-26) also gives overspeed protection should the switch remain

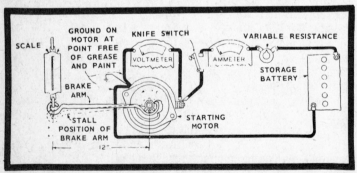

Fig. 10-24. Stall torque test. Multiply scale reading by lever length in feet to obtain torque output. (Courtesy Onan.)

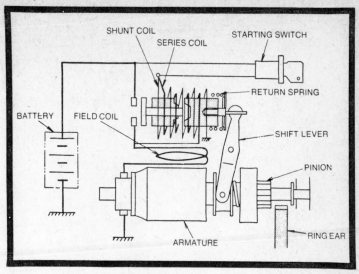

Fig. 10-25. Solenoid internal wiring diagram.

depressed after the engine fires. It functions in conjunction with a direct current generator. Voltage on the relay windings is the difference between generator output and battery terminal voltage. When the engine is cranking, generator output is functionally zero. The relay closes and completes the circuit to the solenoid. When the engine comes up to speed, generator output bucks battery voltage and the relay opens, automatically disengaging the starter. As admirable as such a device is it should not be used continually, since some starter overspeeding will still occur, with detrimental effects to the bearings.

Should the generator circuit open, the starter will be inoperative since the return path for the overspeed relay is through the generator. In an emergency the engine can be started by bridging the overspeed terminals. Of course, the transmission or other loads must be disengaged and, in a vehicle, the handbrake must be engaged.

In addition, you will notice that the solenoid depicted in Fig. 10-26 has two sets of contacts. One set closes first and allows a trickle of current to flow through the resistor (represented by the wavy line above the contact arm) to the motor. The armature barely turns during the engagement phase. But once engaged a trigger is released and the second set of contacts closes, shunting the resistor and applying full battery current to the motor. This circuit, developed by CAV, represents a real improvement over the brutal spin-and-hit

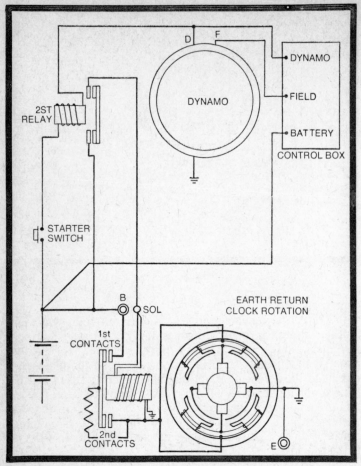

Fig. 10-26. Overspeed relay and solenoid wiring diagram. (Courtesy GM Bedford Diesel.)

action of solenoid-operated and inertia clutches and should result in longer life for all components, including the flywheel ring gear.

Solenoids and relays can best be tested by bridging the large contacts. If the starter works, you know that the component has failed. Relays are sealed units and not repairable. But most solenoids are at least amenable to inspection. Repairs to the series or shunt windings (illustrated in Figs. 10-25 and 10-26) are out of the question unless the circuit has opened at the leads. Contacts may be burnished, and some designs have provision for reversing the copper switch element.

## STARTER DRIVES

Most diesel motors feature positive engagement drives energized by the solenoid. The solenoid may be mounted piggyback on the frame and the pinion moved by means of a pivoted yoke (Fig. 10-25), or it may be coaxial (Fig. 10-7) and the pinion operated directly.

Regardless of mechanical differences between types, all starter drives have these functions:

1. The pinion must be moved laterally on the shaft to engage the flywheel.
2. The pinion must be allowed to disengage when flywheel rpm exceeds pinion rpm.
3. The pinion must be retracted clear of the flywheel when the starter switch is opened.

Figure 10-27 illustrates a typical drive assembly. The pinion moves on a helical thread. Engagement is facilitated by a bevel on the pinion and the ring gear teeth of the flywheel. Extreme wear on either or both profiles will lock the pinion.

The cluch shown employs ramps and rollers. During the motor drive phase the rollers are wedged into the ramps (refer to Fig. 10-28). When the engine catches, the rollers are freed and the clutch overruns. Other clutches employ balls or, in a few cases, ratchets. The spring retracts the drive when the solenoid is deactivated.

Inspect the pinion teeth for excessive wear and chipping. Some battering is normal and does not affect starter operation. The clutch mechanism should be disassembled (if possible), cleaned, and inspected. Inspect the moving parts for wear or deformation, with particular attention to the ratchet teeth and the ramps. Lubricate with Aero Shell 6B or the

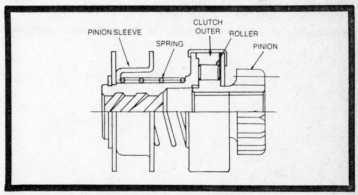

Fig. 10-27. Typical starter drive. (Courtesy Marine Engine Div., Chrysler Corp.)

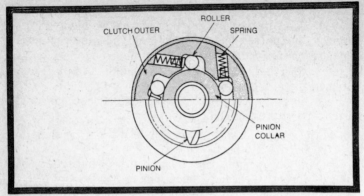

Fig. 10-28. Overrunning clutch. (Courtesy Marine Engine Div., Chrysler Corp.)

equivalent. Sealed drives should be wiped with a solvent-wetted rag. Do not allow solvent to enter the mechanism, since it will dilute the lubricant and cause premature failure. Test the clutch for engagement in one direction of pinion rotation and for disengagement in the other.

Adjustment of the *pinion throw* is important to insure complete and full mesh at the flywheel. Throw is measured between the pinion and the stop ring as shown in Fig. 10-29. Adjustment is by adding or subtracting shims at the solenoid housing (Fig. 10-30), moving the solenoid mounting bolts in their elongated slots, or by turning the yoke pin eccentric.

## CHARGING SYSTEMS

The charging system restores the energy depleted from the battery during cranking and provides power to operate lights and other accessories. It consists of two major components: a generator (or *dynamo*) and a regulator. The

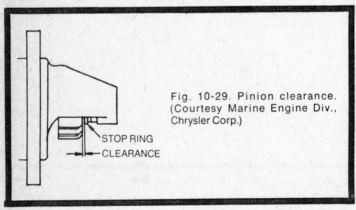

Fig. 10-29. Pinion clearance. (Courtesy Marine Engine Div., Chrysler Corp.)

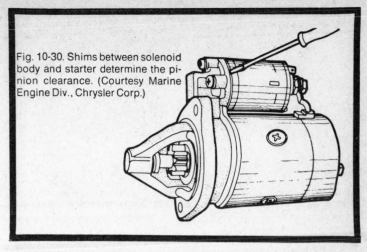

Fig. 10-30. Shims between solenoid body and starter determine the pinion clearance. (Courtesy Marine Engine Div., Chrysler Corp.)

circuit may be monitored by an ammeter or a lamp and is usually fused to protect the generator windings.

### DC Generators

Although on the verge of obsolescence, direct current generators are still fitted to some engines, and thousands of these machines remain in use. Structurally a generator has a kissing cousin resemblance to a starter motor. Major components—frame, armature, brushes, end pieces—are shared by both, although detail differences are not difficult to spot. Most generators have only two field poles in parallel to the armature as opposed to four in series for starter motors. Generator windings are made of finer wire than starter windings, and only two brushes are used.

However, these differences are insignificant insofar as test procedures are concerned. The tests outlined previously apply to generators as well as to starters, with one proviso. The majority of generators are wired in what is known as the *type A configuration*. The fields are insulated from the frame and grounded externally. A few Lucas and vintage American type *B* designs have been encountered with grounded fields (see Fig. 10-31).

To test output of type A machines disconnect the field (F) terminal at the regulator or generator. Output should be zero unless the field windings have shorted to the frame. Next, with all accessories off, jumper the field to ground. Run the engine to speed and observe output. Voltage should treble its normal value, and current output will be the maximum the generator can produce. Make this test quickly: More than a few seconds will melt the armature and field insulation.

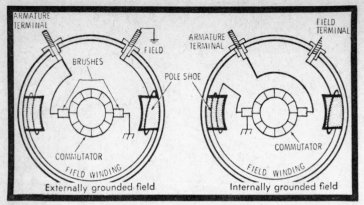

Fig. 10-31. Externally grounded (type A) and internally grounded (type B) fields.

Type B generators are internally grounded. Disconnect the field lead from the armature terminal; there should be no output. With accessories off, momentarily connect the field to the armature (A. ARM. or D); output should jump.

Whenever a new or rebuilt generator is installed it should be polarized; otherwise, the generator may not charge. To polarize type A machines, quickly touch the A lead to the battery (B or BAT) terminal on the voltage regulator. For type B generators touch the field (F) lead to the B terminal.

Belt tension should be adjusted at the generator by slacking the pivot bolts and the generator-arm capscrew. Specifications vary according to generator output and power absorption. Most call for ½ in. of slack between pulley centers with 10 lb of load. Gages are available to take the guesswork out of adjustment and should be used wherever possible. The old-timers used the edge of their hands: An overly tight belt feels solid, and a loose belt feels "dead." A correctly loaded belt is "live " or springy.

## ALTERNATORS

Alternating current generators, or alternators, were introduced in 1960 by Chrysler in the Valiant series of cars. Since then they have come into general use in cars and have spilled over into diesel engine applications. They have several advantages. In the first place alternators are lighter and more compact for the output than are direct current generators. Current output comes on the line early; at idle an alternator will deliver up to 10A, while a generator must be turned at 2000 rpm or so to deliver any appreciable current. This means that the battery can be kept at a constant state of charge regardless of engine speed. Battery life is increased, and all of its stored energy is available. Alternators automatically limit current output, while generators deliver current in rough proportion to rpm. Both, however, require voltage regulation.

The disadvantages of alternators are sometimes exaggerated by mechanics who have not taken the time to understand them. Service is less frequent, although special tools may be needed to remove and install drive sheaves, bearings, and diodes. These tools are initially expensive because of the limited market. Wrong polarity will destroy the diodes and may damage the wiring harness. Observe the polarity when installing a new battery or when using jumper cables. Before connecting a charger to the battery, disconnect the cables. Should the engine be started with the charger in the circuit the regulator may be damaged. Isolate the charging system before any arc welding is done. Do not disconnect the battery or any other wiring while the alternator is turning. And, finally, do not attempt to polarize an alternator. The exercise is fruitless and can destroy diodes.

**Initial Tests**

The charge light should be on with the switch on and the engine stopped. Failure to light indicates an open connection in the bulb itself or in the associated wiring. Most charging-lamp circuits operate by a relay under the voltage regulator cover. Lucas systems employ a separate relay which responds to heat. The easiest way to check either type is to insert a 0−100A ammeter in series with the charging circuit. If the meter shows current and the relay does not close, one can safely assume that it has failed and should be replaced. The Lucas relay can be tested as shown in Fig. 10-32. You will need a voltage divider (or an old-style 12V battery with external cell connections) and 2.2W lamp. Connect clip $A$ to the 12V terminal. The lamp should come on. Leaving the 12V connection in place, connect clip $B$ to the 6V tap. The bulb should burn for 5 sec or so and go out. Move $B$ to the 12V post and hold for no more than 10 sec. Then move it to the 2V (single cell) tap. The bulb should come on within 5 sec. These units do not have computer-like precision, and some variation can be expected between them. But the test results should roughly correlate with the test procedure. Do not attempt to repair a suspect relay.

Test the alternator output against the meter on the engine or by inserting a test meter in series between the $B$ terminal and battery as shown in Fig. 10-32. Voltage is monitored with a meter in parallel with the charging circuit. Discharge the battery by switching on the lights and other accessories.

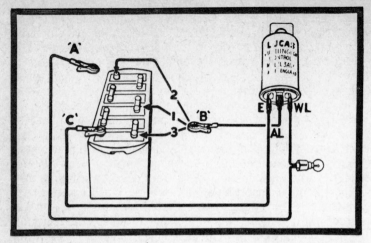

Fig. 10-32. Testing Lucas charging lamp relay. To distinguish these relays from turn-signal flashers, Lucas has coded them green. (Courtesy GM Bedford Diesel.)

Connect a rheostat or carbon pile across the battery for a controlled discharge. (Without this tool you will be reduced to guessing about alternator condition.) With the load set at zero, start the engine and operate at approximately midthrottle. Apply the load until the alternator produces its full rated output. If necessary open the throttle wider. An output 2—6A below rating often means an open diode. Ten amps or so below rating usually means a shorted diode. The alternator may give further evidence of a diode failure by whining like a wounded banshee.

The voltage should be 18—20V above the nominal battery voltage under normal service conditions. It may be higher by virtue of automatic temperature compensation in cold weather.

Assuming that the output is below specs, the next step is to isolate the alternator from the regulator. Disconnect the field (F or FD) terminal from the regulator and ground it to the block. Load the circuit with a carbon pile to limit the voltage output. Run the engine at idle. In this test we have dispensed with the regulator and are protecting the alternator windings with carbon pile. No appreciable output differences between this and the previous tests means that the regulator is doing its job. A large difference would indicate that the regulator is defective.

Late-production alternators often have integrated regulators built into the slipring end of the unit. Most have a provision for segregating alternator output from the regulator so that "raw" outputs may be measured. The Delcotron features a shorting tab. A screwdriver is inserted into an access hole in the back of the housing (Fig. 10-33); contact between the housing and the tab shorts the fields.

*Note*: The tab is within ¾ in. of the casting. Do not insert a screwdriver more than 1 in. into the casting.

## Bench Testing

Disconnect the battery and remove the alternator from the engine at the pivot and belt-tensioning bracket. Three typical alternators are shown in exploded view in Figs. 10-34 through 10-36. Note that these two English and one American design are more alike than different.

Remove the drive pulley. A special tool may be needed on some of the automotive derivations. Hold the fan with a screwdriver and turn the fan nut counterclockwise. Tap the sheave and fan off the shaft with a mallet. Remove the throughbolts and separate the end shields.

Inspect the brushes for wear. Some manufacturers thoughtfully provide a wear limit line on the brushes (Fig. 10-37). Clean the holders with Freon or some other nonpetroleum-based solvent and check the brushes for ease of movement. File lightly if they appear to bind. The sliprings should be miked for wear and eccentricity. Ten to twelve thousandths should be considered the limit (Fig. 10-38). Sliprings are usually, but not always, integral with the rotor. Removable rings are chiseled off and new ones pressed into place. Fixed rings can be restored to concentricity with light machining.

Determine the condition of the rotor insulation with a 110V test lamp (Fig. 10-39). The sliprings and their associated windings should be insulated from the shaft and pole pieces. If you have access to an accurate, low-range ohmmeter, test for

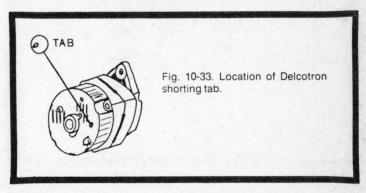

Fig. 10-33. Location of Delcotron shorting tab.

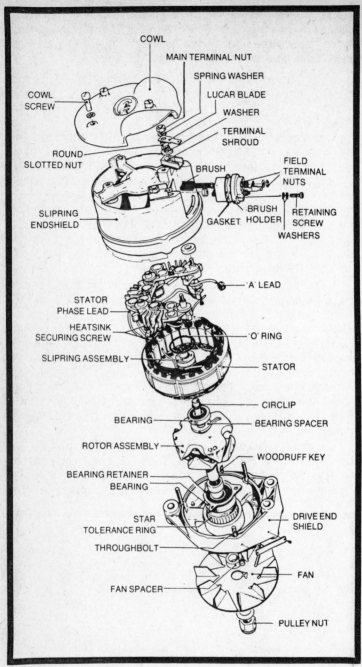

Fig. 10-34. CAV AC5-24 alternator. (Courtesy GM Bedford Diesel.)

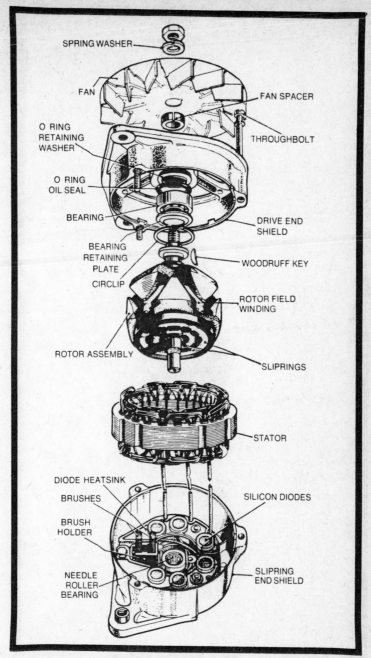

Fig. 10-35. Lucas 10-AC or 11-AC alternator. (Courtesy GM Bedford Diesel.)

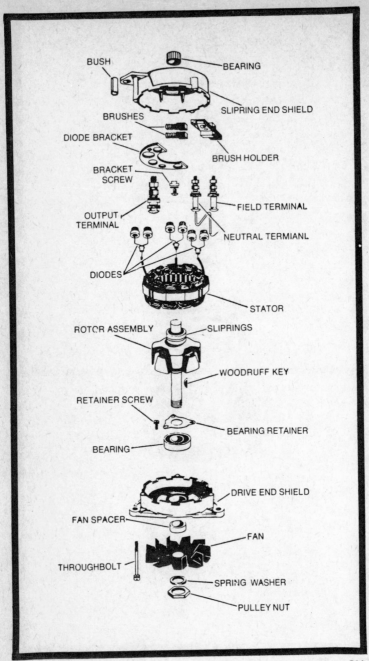

Fig. 10-36. Prestolite CAB-1235 or CAB-1245 alternator. (Courtesy GM Bedford Diesel.)

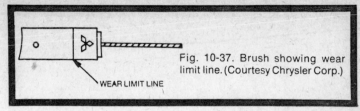

Fig. 10-37. Brush showing wear limit line. (Courtesy Chrysler Corp.)

WEAR LIMIT LINE

continuity between sliprings. The resistance may lead one to suspect a partial open; less could mean an intracoil short.

The stator consists of three distinct and independent windings whose outputs are 120° apart. It is possible for one winding to fail without noticeably affecting the others. Peak

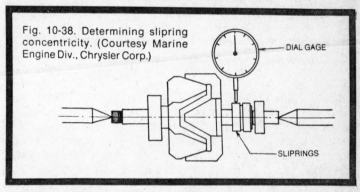

Fig. 10-38. Determining slipring concentricity. (Courtesy Marine Engine Div., Chrysler Corp.)

DIAL GAGE

SLIPRINGS

alternator output will, of course, be reduced by one third. Disconnect the three leads going to the stator windings. Many European machines have these leads soldered, while American designs generally have terminal lugs. When unsoldering, be extremely careful not to overheat the diodes.

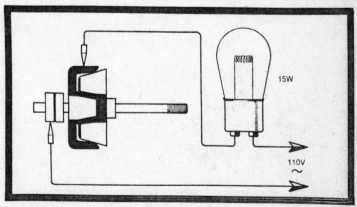

15W

110V
~

Fig. 10-39. Checking rotor insulation. (Courtesy GM Bedford Diesel.)

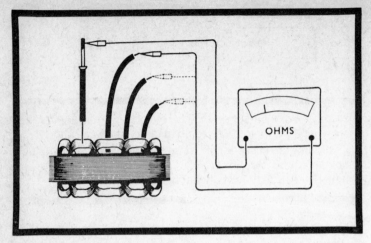

Fig. 10-40. Comparison test between stator windings. (Courtesy GM Bedford Diesel.)

Exposure to more than 300°F will upset their crystalline structure. Test each winding for resistance. Connect a low-range ohmmeter between the neutral lead and each of three winding leads as shown in Fig. 10-40. Resistance will be quite low—on the order of 5 or 6 ohms—and becomes critical when one group of windings gives a different reading than the others.

Test the stator insulation with a 110V lamp connected as shown in Fig. 10-41. There should be no continuity between the laminations and windings.

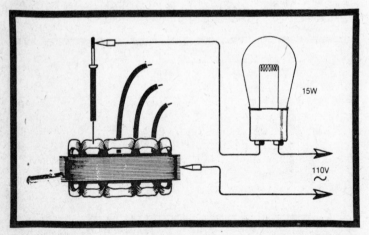

Fig. 10-41. Comparison test between stator windings. (Courtesy GM Bedford Diesel.)

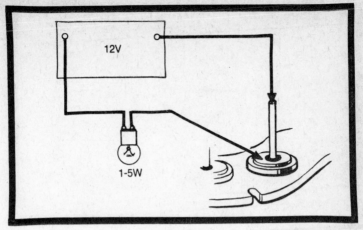

Fig. 10-42. Diode testing. (Courtesy GM Bedford Diesel.)

The next step is to check the diodes (Fig. 10-42). You may already have had some evidence of diode trouble in the form of alternator whine or blackened varnish on the stator coils. The diodes must be tested with an ohmmeter or a test lamp of the same voltage as generator output.

Test each diode by connecting the test leads and then reversing their polarity. The lamp should light in one polarity and go out in the other. Failure to light at all means an open diode; continuous burning means the diode has shorted. In either case it must be replaced. You may use a low-voltage ohmmeter in lieu of a lamp. Expect high (but not infinite) resistance with one connection, and low (but not zero) resistance when the two leads are reversed.

To simply service and limit the need for special tools, some manufacturers package mounting brackets with their diodes. The bracket is a heatsink and must be in intimate contact with the diode case. Other manufacturers take the more traditional approach and supply individual diodes, which must be pressed (not hammered) into their sinks. K-D Tools makes a complete line of diode removal and installation aids, including heatsink supports and diode arbors of various diameters. Figure 10-43 shows a typical installation with an unsupported heatsink. Other designs may require support.

Soldering the connections is very critical. Should the internal temperature reach 300°F the diode will be ruined. Use a 150W or smaller iron and place a thermal shunt between the soldered joint and the diode (Fig. 10-44). The shunt may be in the form of a pair of needle-nosed pliers or copper alligator clips. In some instances there may not be room to shunt the

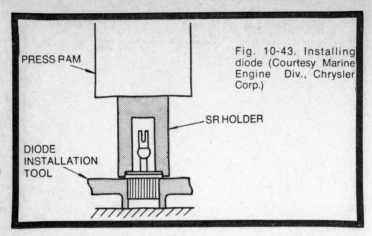

PRESS RAM

SR HOLDER

DIODE INSTALLATION TOOL

heat load between the diode and joint (Fig. 10-44B). It is only necessary to twist the leads enough to hold them while the solder is liquid. Work quickly and use only enough solder to flow between the leads. More solder merely increases the thermal load and increases the chances that the diode will be ruined.

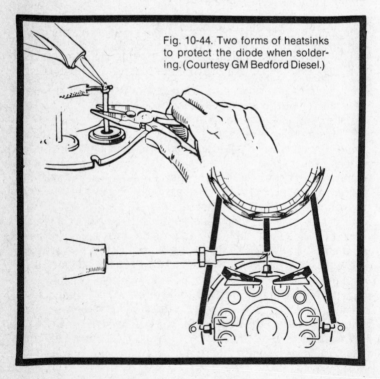

Alternator bearings are sealed needle and ball types. They are not to be disturbed unless noisy or rough. Then bearings are pressed off and new ones installed with the numbered end toward the arbor.

## DC REGULATOR SYSTEMS

Almost all direct current generator systems are fitted with 3-element vibrating-reed regulators. A few installations have been updated with transistorized regulators.

Each element consists of a relay with windings, point contacts, and (in most cases) resistors in shunt. The *cutout relay* is double-wound. One winding is in parallel and the other in series. When the generator starts to turn, current flows through the parallel winding and closes the cutout relay contacts. Additional current takes the series route. Both add together to energize the coil and keep the contacts closed. When the engine is shut down or turning at low speed, current flows from the battery through the cutout relay and to the generator. This state of affairs cannot be allowed to continue, since the battery would discharge itself through the generator ground brush. As the differential between output and stored energy in the battery increases, more current flows through the relay windings. Because of the lay of the two windings, current from the battery generates a bucking, or opposing, field in the series winding. This field tends to cancel the field generated by the parallel winding. The spring-loaded contacts open, isolating the generator from the battery. You can trace this circuit in Fig. 10-45.

The voltage-sensitive winding is in parallel with the output and can be identified by its fine-gage windings. Normally the contacts are closed. As the generator comes up to speed, voltage increases until the voltage-sensitive winding develops

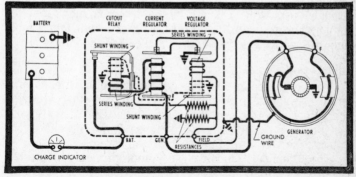

Fig. 10-45. Dc charging system, type A.

a strong enough field to open the contacts. Current from the fields is diverted to a resistor which drops the excess voltage.

The current-sensitive winding consists of a few turns of heavy wire. It is in series with the output and receives all current leaving the armature. At a certain preset value current in the series windings causes the contacts to open, diverting field current through a limiting resistor. The reduced field current results in a reduced generator output.

Mechanical regulators have been subject to much refinement over the years. All but the most primitive have some means of temperature compensation. As temperatures increase the battery becomes more chemically active, and charging efficiency goes up. Less input is needed. The regulator senses temperature in any of several ways. Some manufacturers employ a bimetallic spring whose tension responds to temperature changes; others employ a magnetic bypass on the coils to bleed the field and reduce the pull on the coil armature under high temperatures. Others employ either of these methods in combination with temperature-sensitive resistors.

Dual contacts may be fitted to the voltage regulator in the case of engines which have a broad rpm band. The second set of contacts comes into play past the operating point of the first. Instead of diverting field current to a resistor, these contacts open the field circuit completely, dropping generator output to almost zero.

## Regulator Servicing

A regulator is a highly sensitive device and should not be opened for casual inspection or tinkering. The care which goes into their construction approaches or matches that lavished on injector pumps. For example, the final adjustment room at DelcoRemy is supplied by four conveyor belts entering through each wall. The air in the room is monitored for humidity, temperature, and dust. If more than four people enter the room an alarm sounds and the lines shut down.

## Adjustment

The adjustments may be accomplished by means of a screw, bending the stationary contacts, or by means of moving the hinges in elogated mounting slots. Check with your distributor for detailed instructions. Some of the more elaborate regulators require special instruments.

The cutout relay should close at approximately 13V on 12V systems and 26.8V on 24V systems. Connect a voltmeter between the output (A or D terminal) and ground. Put a 10A load on the battery and slowly run the engine up to speed. The meter needle will flick when the contacts close.

The voltage regulator is usually adjusted from a voltage reading taken between the battery terminals. A more accurate method is to remove the regulator cover and isolate the battery by inserting a piece of paper between the cutout contacts. With the battery isolated, the generator is running on an open circuit and can develop high voltages which could cause overheating of the voltage-limiting coil. So work carefully. With a voltmeter connected between the generator output and ground, slowly open the throttle. The needle should flick at the prescribed voltage, which should be approximately 15V at 68°F, or 28V on 24V systems.

To make the current adjustment, connect an ammeter in series with the generator output.

Manufacturers' suggested procedures vary. Some prefer the engine to be run to a predetermined speed with the voltage contacts shunted. Others suggest a load be put on the system. You will need detailed accurate specifications, which cannot be included in a book of this type.

Oxidized contacts may be burnished with a riffle file. *Do not use sandpaper or emery cloth.* Contact gap, air gap (the clearance between the coil cores and the movable armatures), and yoke gap are adjustable (see Fig. 10-46). Inspect the contact points, springs, windings, and resistors for telltale signs of overheating. Check the regulator ground connection.

Before discarding a defective regulator, attempt to discover why it failed; otherwise, the new one may fall victim

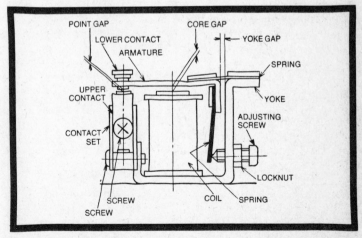

Fig. 10-46. A typical relay and its adjustment points. (Courtesy Marine Engine Div., Chrysler Corp.)

to the same malfunction. Burnt points or discolored springs on the voltage or current relay mean high resistance in the charging circuit or a bad regulator ground. A burnt coil points to the same problem. If the damage is localized to the current winding points, you can expect to find a short somewhere in the system—maybe in the battery. Burnt points or discolored springs on the cutout indicate that the generator was not polarized.

The regulator should be mounted per the manufacturer's instructions and well clear of engine compartment heat. If the unit "machine-guns" on starting, shut down immediately and polarize the generator.

## AC REGULATOR SYSTEMS

Most regulators intended for use with alternators control the voltage. The upper limit of current output is determined by voltage. Operation is similar to that of the direct current type. Figure 10-47 illustrates a 2-contact voltage-regulating relay used in conjunction with the Hitachi 300W alternator. The strength of the magnetic field created by the voltage coil is

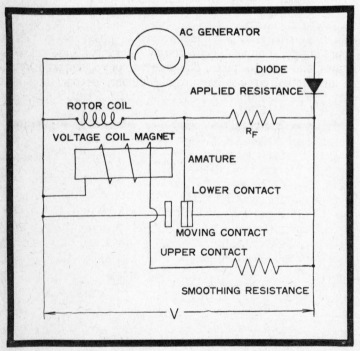

Fig. 10-74. Voltage regulator in alternator circuit. (Courtesy Marine Engine Div., Chrysler Corp.)

proportional to the current through it, which in turn is a function of voltage. At a preset voltage the moving contact is pulled away from the lower contact. Field (rotor coil) current is limited by passing through the applied resistance $R_F$. At very high rotational speeds more voltage will induced in the circuit, and the movable contact will be pulled to the upper contact, shorting the field and reducing output.

Like so many of these regulators which were originally developed for automobiles, this one includes a charging lamp relay. The lamp burns when the contacts are closed. When the alternator voltage rises to 70−80% of normal output, the contacts open, interrupting the charging-lamp circuit.

These relays are serviced as described in the preceding section.

## SOLID-STATE REGULATORS

Transistorized regulators were pioneered by small-engine manufacturers and have found their way into automobile and diesel applications. They are capable of exceedingly fine regulation, partially because there are no moving parts. Durability is exceptional. On the other hand any internal malfunction generally means that the unit must be replaced. As a rule, no repairs are possible. Failure can occur because of manufacturing error (this usually shows up in the first few hours of operation and is covered by warranty), high current draws, and voltage spikes. The presence of voltage transients in automotive circuits has given engineers second thoughts about these regulators in autos and light trucks. Diesel wiring is usually quite simple and less subject to transients.

The mechanic must be particularly alert when working with transistorized circuits. The cautions which apply to alternator diodes apply with more force to regulators if only because regulators are more expensive. Do not introduce stray voltages, cross connections, reverse battery polarity, or open connections while the engine is running.

The regulator may be integral with the alternator or may be contained in a separate box. In general, no adjustment is possible; however, the Lucas 4TR has a voltage adjustment on its bottom, hidden under a dab of sealant.

## BATTERIES

The battery has three functions: provide energy for the starting motor; stabilize voltages in the charging system; and, for limited periods, provide energy for the accessories in the event of charging-circuit failure. Because it is in a constant state of chemical activity and is affected by temperature changes, aging, humidity, and current demands, the battery requires more attention than any other component in the electrical system.

## Battery Ratings

Starting a diesel engine puts a heavy drain on the battery, especially in cold climates. One should purchase the best quality and the largest capacity practical. The physical size of the battery is coded by its *group number*. The group number has only an indirect bearing on electrical capacity but does assure that replacements will fit the original brackets. In some instances a larger capacity battery may require going to another group number. Expect to modify the bracket and possibly to replace one or more cables.

The traditional measure of a battery's ability to do work is its *ampere-hour* (A-hr) *capacity*. The battery is discharged at a constant rate for 20 hr so that the potential of each cell drops to 1.75V. A battery that will deliver 6A over the 20 hr period is rated at 120 A-hr (6A × 20 hr). You will find this rating stamped on replacement batteries or in the specifications.

Like all rating systems, the ampere-hour rating is best thought of as a yardstick for comparison between batteries. It has absolute validity only in terms of the original test. For example, a 120 A-hr battery will not deliver 120A for 1 hr, nor will it deliver 1200A for 6 min.

*Cranking-power* tests are more meaningful since they take into account the power loss which lead−acid batteries suffer in cold weather. At room temperature the battery develops its best power; power output falls off dramatically around 0°F. At the same time, the engine becomes progressively more difficult to crank and more reluctant to start. Several cranking-power tests are in use.

*Zero cranking power* is a hybrid measurement expressed in volts and minutes. The battery is chilled to 0°F; depending upon battery size, a 150 or 300A load is applied. After 5 sec the voltage is read for the first part of the rating. Discharge continues until the terminal voltage drops to 5V. The time in minutes between full charge and effective exhaustion is the second digit in the rating. The higher these two numbers are for batteries in the same load class, the better.

The *cold cranking performance* rating is determined by lowering the battery temperature to 0°F (or, in some instances, 20°F) and discharging for 30 sec at such a rate that the voltage drops below 1.2V per cell. This is the most accepted of all cold weather ratings and has become standard in specification sheets.

## Battery Tests

As the battery discharges, some of the sulfuric acid in the electrolyte decomposes into water. The strength of the electrolyte in the individual cells is a reliable index of the state of charge. There are several ways to determine acidity, but long ago technicians fixed upon the measurement of specific gravity as the simplest and most reliable.

The instrument used is called a *hydrometer* (Fig. 10-48). It consists of a rubber bulb, a barrel, and a float with a graduated tang. The graduations are in terms of specific gravity. Water is assigned a specific gravity of 1. Pure sulfuric acid is 1.83 times heavier than water and thus has a specific gravity of 1.83. The height of the float tang above the liquid level is a function of fluid density, or specific gravity. The battery is said to be fully charged when the specific gravity is between 1.250 and 1.280.

An accurate hydrometer test takes some doing. The battery should be tested prior to starting and after the engine has run on its normal cycle. For example, if the engine is shut down overnight, the test should be made in the morning, before the first start. Water should be added several operating days before the test to insure good mixing. Otherwise the readings may be deceptively low.

Use a hydrometer reserved for battery testing. Specifically, do not use one which has been used as an antifreeze tester. Trace quantities of ethylene glycol will shorten the battery's life.

Place the hydrometer tip above the plates: Contact with them may distort the plates enough to short the cell. Draw in a generous supply of electrolyte and hold the hydrometer vertically. You may have to tap the side of the barrel with your

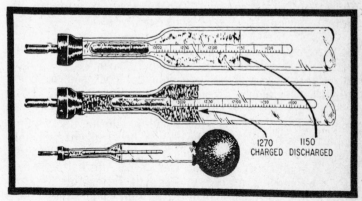

1270 CHARGED    1150 DISCHARGED

Fig. 10-48. Hydrometer.

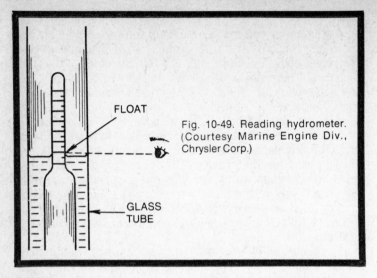

FLOAT

Fig. 10-49. Reading hydrometer. (Courtesy Marine Engine Div., Chrysler Corp.)

GLASS TUBE

fingernail to jar the float loose. Holding the hydrometer at eye level, take a reading across the fluid level. Do not be misled by the meniscus (concave surface, Fig. 10-49) of the fluid.

American hydrometers are calibrated to be accurate at 80°F. For each 10°F above 80°F, add 4 points (0.004) to the reading; conversely, for each 10°F below the standard, subtract 4 points. The standard temperature for European and Japanese hydrometers is 20C, or 68°F. For each 10°C increase add 7 points (0.007); subtract a like amount for each 10°C decrease. The more elaborate hydrometers have a built-in thermometer and correction scale.

All cells should read within 50 points (0.050) of each other. Greater variation is a sign of abnormality and may be grounds for discarding the battery. The relationship between specific gravity and state of charge is shown in Fig. 10-50.

The hydrometer test is important, but by no means definitive. The state of charge is only indirectly related to the actual output of the battery. Chemically the battery may have full potential, but unless this potential passes through the straps and terminals, it is of little use.

Perhaps the single most reliable test is to load the battery with a rheostat or carbon pile while monitoring the terminal voltage. The battery should be brought up to full charge before the test. The current draw should be adjusted to equal three times the ampere-hour rating. Thus, a 120 A-hr battery would be discharged at a rate of 360A. Continue the test for 15 sec and observe the terminal voltage. At no time should the voltage drop below 9.5V.

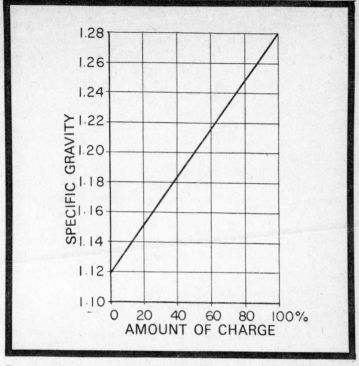

Fig. 10-50. Relationship between state of charge and specific gravity. (Courtesy Marine Engine Div., Chrysler Corp.)

In this test, sometimes called the *battery capacity* test, we used voltage as the telltale. But without a load, terminal voltage is meaningless. The voltage remains almost constant from full charge to exhaustion.

### Battery Maintenance

The first order of business is to keep the electrolyte level above the plates and well into the reserve space below the filler cap recesses. Use distilled water. Tap water may be harmful, particularly if it has iron in it.

Inspect the case for cracks, acid seepage, and (in the old asphalt-based cases) softening. Periodically remove the cable clamps and scrape them and the battery terminals. Look closely at the bond between the cables and clamp. The best and most reliable cables have forged clamps, solder-dipped for conductivity. Replace spring clip and other clever designs with standard boltup clamps sweated to the cable ends. After scraping and tightening, coat the terminals and clamps with grease to provide some protection from oxidation.

The battery case should be wiped clean with a damp rag. Dirt, spilled battery acid, and water are conductive and promote self-discharge. Accumulated deposits can be cleaned and neutralized with a solution of baking soda, water, and detergent. Do not allow any of the solution to enter the cells, where it would dilute the electrolyte. Rinse with clear water and wipe dry.

### Charging

Any type of battery charger may be used: selenium rectifier, tungar rectifier, or, reaching way back, mercury arc rectifier. Current and voltage should be monitored and there should be a provision for control. When charging multiple batteries from a single output, connect the batteries in series as shown in Fig. 10-51.

Batteries give off hydrogen gas, particularly as they approach full charge. When mixed with oxygen, hydrogen is explosive. Observe these safety precautions:

- Remove all filler caps (to prevent pressure rise should the caps be clogged).
- Charge in a well ventilated place remote from open flames or heat.
- Connect the charger leads before turning the machine on. Switch the machine off before disconnecting the leads.

In no case should the electrolyte temperature be allowed to exceed 115°F. If your charger does not have a thermostatic

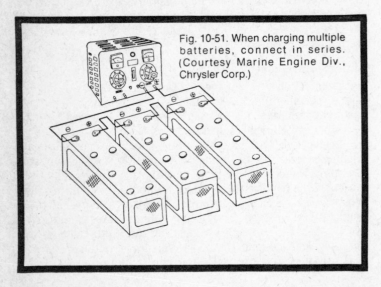

Fig. 10-51. When charging multiple batteries, connect in series. (Courtesy Marine Engine Div., Chrysler Corp.)

control, keep track of the temperature with an ordinary thermometer.

Batteries may be charged by any of three methods. *Constant-current* charging is by far the most popular. The charging current is limited to one-tenth of the ampere-hour rating of the battery. Thus a 120 A-hr battery would be charged at 12A. Specific gravity and no-load terminal voltage are checked at 30 min intervals. The battery may be said to be fully charged when both values peak (specific gravity 1.127−1.129, voltage 15−16.2V) and hold constant for three cranking intervals.

A *quick charge*, also known as a *booster* or *hotshot*, can bring a battery back to life in a few minutes. The procedure is not recommended in any situation short of an emergency, since the high-power boost will raise the electrolyte temperature and may cause the plates to buckle. Disconnect the battery cables to isolate the generator or alternator if such a charge is given the battery while it is in place.

A constant-voltage charge may be thought of as a compromise between the hotshot and the leisurely constant-current charge. The idea is to apply a charge by keeping charger voltage 2.2−2.4V higher than terminal voltage. Initially the rate of charge is quite high; it tapers off as the battery approaches capacity.

**Battery Hookups**

One of the most frequently undertaken field modifications is to add additional batteries. The additional capacity make starts easier in extremely cold weather and adds reliability to the system.

To increase capacity add one or more batteries in parallel: negative post connected to negative post, positive post to positive post. The voltage will not be affected, but the capacity will be the sum of all parallel batteries.

Connecting in series—negative to positive, positive to negative—adds voltage without changing capacity. Two 6V batteries can be connected in series to make a physically large 12V battery.

Cable size is critical since the length and cross-sectional area determine resistance. Engineers at International Harvester have developed the following recommendations which may be applied to most small, high-speed diesel engines:

1. Use cables with integral terminal lugs.
2. Use only rosin or other noncorrosive-flux solder.

3. Terminal lugs must be stacked squarely on the terminals. Haphazard stacking of lugs should be avoided.
4. Where the frame is used as a ground return, it must be measured and this distance added to the cable length to determine the total length of the system. Each point of connection with the frame must be scraped clean and tinned with solder. There would be no point of resistance in the frame such as a riveted joint. Such joints should be bridged with a heavy copper strap.
5. Pay particular attention to engine—frame grounds. If the engine is mounted in rubber it is, of course, electrically isolated from the frame.
6. Check the resistance of the total circuit by the voltage drop method or by means of Ohm's law ($R = E/I$).
7. Use this table as an approximate guide to cable size for standard-duty cranking motors.

| System Voltage | Maximum Resistance | Cable Size and Length |
|---|---|---|
| 12V | 0.0012 ohm | Less than 105 in., No. 0 |
| | | 105 to 132 in., No. 00 |
| | | 132 to 162 in., No. 000 |
| | | 162 to 212 in., No. 0000 in parallel |
| 24V | 0.0020 ohm | Less than 188 in., No. 0 |
| | | 188 to 237 in., No. 00 |
| | | 237 to 300 in., No. 0000 |
| | | 300 to 380 in., No. 0000, or two No. 0 in parallel |

8. Suggested battery capacity varies with system voltage (the higher the voltage, the less capacity needed for any given application), engine displacement, compression ratio, ambient air temperature, and degree of exposure. Generalizations are difficult to make, but typically a 300 CID engine with a 12V standard-duty starter requires a 700A battery for winter operation. This figure is based on SAE J-5371 specifications and refers to the 30 sec output of a chilled battery. At 0°F capacity should be increased to 900A. The International Harvester 414 CID engine requires 1150, or 1400A at 0°F. Battery capacity needs roughly parallel engine displacement figures, with some flattening of the curve for the larger and easier-to-start units. In extremely cold weather—below −10°F—capacity should be increased by 50% or the batteries heated.

274

# Cooling Systems

**11**

No engine converts all, or even most, of the heat latent in the fuel to useful work. The diesel is the best of them and still wastes about 60% of the heat input. Some of this waste goes out the exhaust, and the rest is absorbed in the cooling system either directly or indirectly by means of the lube oil. Without some form of cooling the engine would quickly self-destruct. The gas temperature in the combustion chamber can exceed 5000°F under normal operation, and gets even hotter when the engine is lugged.

Small diesel engines are air- or liquid-cooled. Air-cooled engines are generally limited to under 40 hp, although Hatz makes a 4-cylinder, 245 CID model developing 80 hp. The only customer for larger air-cooled engines is the military.

## AIR COOLING

The major advantage of air cooling is the simplicity and light weight of such a system. There are no pumps, radiators, hoses, oil coolers, and the like to fail and add pounds of dead weight. On the other hand, air-cooled engines are noisy—the same fins which radiate heat also radiate sound. And air-cooled engines do not have the precise temperature regulation which we associate with liquid cooling. Air-cooled engines reach operating temperature quickly, which is all to the good; but their temperature tends to fluctuate with load and rpm. The flywheel fan pumps more air than is needed at high speeds, and less than is needed when the engine bogs. Of course, this objection does not apply when the load is fixed or when it varies well within the governor's capabilities.

Figure 11-1 shows a typical air-cooled engine. The flywheel has impeller blades cast into its rim. Air moves through the guard and is flung outwards against the shroud and around the cylinders. You may run into an application with reversed air

275

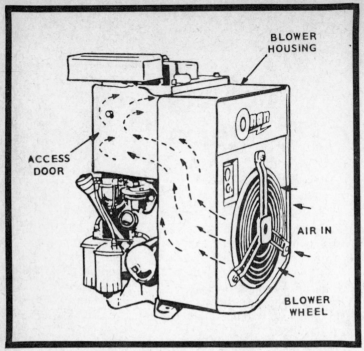

Fig. 11-1. Typical air flow pattern. (Courtesy Onan.)

flow. Some generator sets employ the flywheel fan in suction so that the generator gets the cool breeze first.

Air cooling is not entirely maintenance-free. The fins should be brushed and blown clean periodically, and the shrouds must be kept in good repair. Capscrews which have vibrated loose can be doctored with a dab of Permatex Lock-Nut or the equivalent. The lube oil level is critical in any engine, and especially in these air pumpers. The lube is a heatsink, and the level may drop more quickly than you expect if you are not accustomed to working with air-cooled engines. Because the engines run hotter than the liquid-cooled types, the cold fit of the parts is a trifle loose. Consequently, oil consumption may be slightly higher, although, in theory, everything should be back to normal when operating temperature is reached. For improved temperature control a thermostatically controlled shutter is often employed in this type of system to modulate the air flow (Fig. 11-2). The shutter is opened further by the actuating mechanism in Fig. 11-3 as more cooling is called for.

The shutter-opening temperature is not adjustable, but to insure complete opening, the power element plunger must

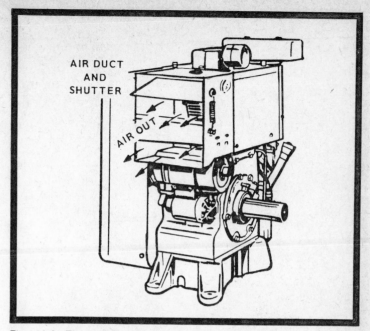

AIR DUCT
AND
SHUTTER

AIR OUT

Fig. 11-2. Thermostatic shutter for improved temperature control.
(Courtesy Onan.)

contact the shutter roll pin at room temperature. The power
element screws should be loosened and the element slid in its
mounts until positive contact is made. Failure to open is a
serious matter and can quickly put you in the market for a new
engine. Check the shutters for freedom of movement and
check the Vernatherm element. Checking the element can be
done by heating it. At about 120°F the plunger should start to
move out of its housing; at 140° or so the plunger should be
extended to its maximum, which is about 0.200 in.

Failure to close completely will rob the engine of power
and waste fuel. If the nylon bushings which support the shutter
pivots are worn or binding, replace them. Remove the shutters
and withdraw the pivot shaft. Install new bearings from the
inside of the housing. The large surface serves as a thrust
bearing and should be on the inside. End-thrust clearance
should be no more than $^1/_{32}$ in. One may assume that the
actuating rod is in adjustment, since it was set at the factory,
and that it will not come out of adjustment unless the housing
is warped or otherwise damaged.

As reliable as this Onan system is, it is not perfect. As a
fail-safe you should install a thermostatic switch, available
from Onan, in series with the solenoid cutoff. The switch is

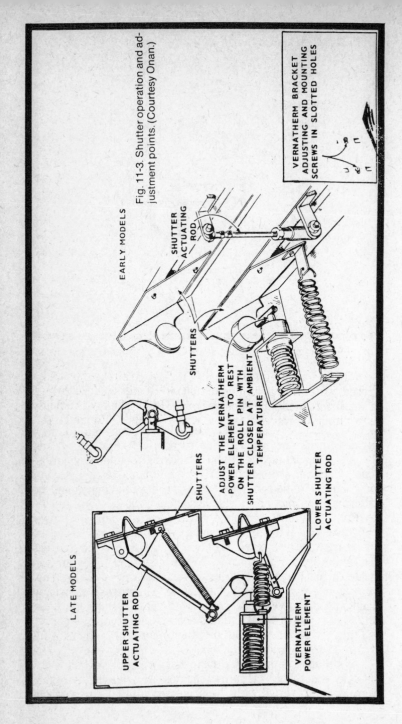

EARLY MODELS

SHUTTER ACTUATING ROD

SHUTTERS

VERNATHERM BRACKET ADJUSTING AND MOUNTING SCREWS IN SLOTTED HOLES

Fig. 11-3. Shutter operation and adjustment points. (Courtesy Onan.)

LATE MODELS

SHUTTERS

ADJUST THE VERNATHERM POWER ELEMENT TO REST ON THE ROLL PIN WITH SHUTTER CLOSED AT AMBIENT TEMPERATURE

UPPER SHUTTER ACTUATING ROD

LOWER SHUTTER ACTUATING ROD

VERNATHERM POWER ELEMENT

278

normally closed and opens at 240°F, cutting off the fuel supply. It closes again at about 195°F. To test the thermostatic switch, heat it while monitoring continuity with an ohmmeter across the terminals.

## LIQUID COOLING

Most diesel engines are liquid-cooled. The additional weight and potential for trouble is not an overriding concern in most applications and is compensated for by good control over internal temperatures, relative silence, and ease of manufacture.

The major parts of a liquid-cooled system are the radiator, which dumps heat collected by the coolant into the atmosphere (in a sense, all engines are air-cooled); the circulation pump; the thermostat, which traps some of the coolant in the head for quicker warmups; the water jacket surrounding the upper engine; and assorted hoses. Figure 11-4 shows a typical layout as used in a vehicle. Stationary and larger vehicular engines may include such refinements as oil and transmission coolers. Marine engines forego the radiator and circulate raw water or, sometimes, fresh water through a heat exchanger.

### Coolant Circuits

The basic circuit is shown in Fig. 11-5. With the thermostat closed, water flow is limited to that which can squeeze through a small port in the thermostat body. The engine heats quickly, but without any dead water areas which could boil. At a predetermined temperature the thermostat opens and circulation is unimpeded.

Air is the enemy of cooling systems and must be kept out of the coolant supply. The usual way to do this is to provide overflow tanks (9 in Fig. 11-4) and internal baffles. The problem is more serious with diesel engines than with gasoline types because diesels tend, as they get older, to leak compression into the coolant. Other sources of air are the pump seals and splash entrapment. Besides reducing the efficiency of the system (air is 3500 times less efficient than water in terms of heat removal), air can damage the pump impeller and cause the coolant to erupt out of the overflow. In the system in Fig. 11-6 the thermostat is shunted by a deaeration line. The line picks up coolant at the highest point in the engine and, hopefully, air as well. When the thermostat opens, coolant still flows through the line because of a built-in restriction in the thermostat body.

The full deaeration system is an elaboration of the one just discussed. As you can see from Fig. 11-7, the thermostat blocks

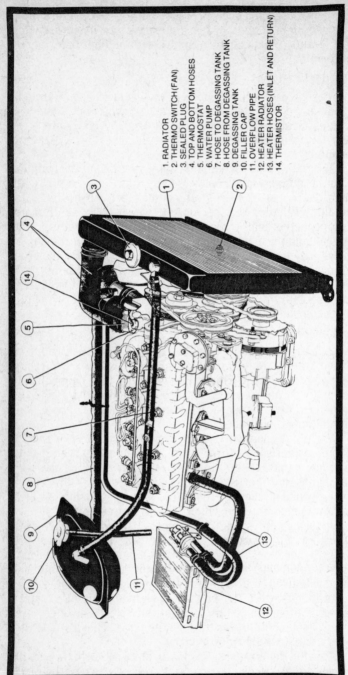

1. RADIATOR
2. THERMO SWITCH (FAN)
3. SEALED PLUG
4. TOP AND BOTTOM HOSES
5. THERMOSTAT
6. WATER PUMP
7. HOSE TO DEGASSING TANK
8. HOSE FROM DEGASSING TANK
9. DEGASSING TANK
10. FILLER CAP
11. OVERFLOW PIPE
12. HEATER RADIATOR
13. HEATER HOSES (INLET AND RETURN)
14. THERMISTOR

Fig. 11-4. Cooling-system diagram. (Courtesy Peugeot.)

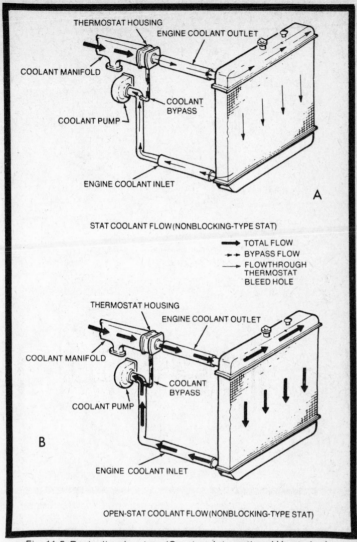

THERMOSTAT HOUSING
ENGINE COOLANT OUTLET
COOLANT MANIFOLD
COOLANT BYPASS
COOLANT PUMP
ENGINE COOLANT INLET
A

STAT COOLANT FLOW (NONBLOCKING-TYPE STAT)

→ TOTAL FLOW
⇢ BYPASS FLOW
→ FLOWTHROUGH THERMOSTAT BLEED HOLE

THERMOSTAT HOUSING
ENGINE COOLANT OUTLET
COOLANT MANIFOLD
COOLANT BYPASS
COOLANT PUMP
B
ENGINE COOLANT INLET

OPEN-STAT COOLANT FLOW (NONBLOCKING-TYPE STAT)

Fig. 11-5. Basic diesel system. (Courtesy International Harvester.)

the radiator and directs the coolant back to the engine. The block and head have the benefit of full circulation at all times. The deaeration line picks up air and separates it from the water in the header tank. When the thermostat opens, the deaeration line works in conjunction with a standpipe. This system has some real advantages, especially in cold climates. The engine warms up faster and, since the radiator cuts in and out of the circuit, the coolant can be at much higher

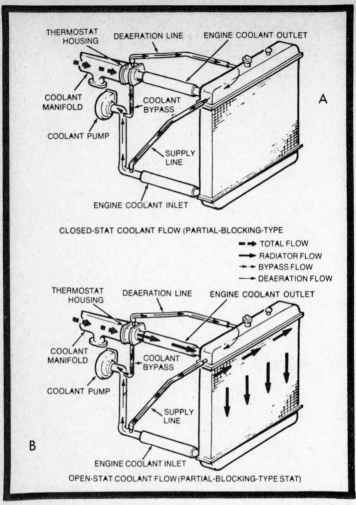

THERMOSTAT HOUSING  DEAERATION LINE  ENGINE COOLANT OUTLET

COOLANT MANIFOLD

COOLANT BYPASS

COOLANT PUMP

SUPPLY LINE

A

ENGINE COOLANT INLET

CLOSED-STAT COOLANT FLOW (PARTIAL-BLOCKING-TYPE

- ▪▪➤ TOTAL FLOW
- ——➤ RADIATOR FLOW
- –•–➤ BYPASS FLOW
- ——➤ DEAERATION FLOW

THERMOSTAT HOUSING  DEAERATION LINE  ENGINE COOLANT OUTLET

COOLANT MANIFOLD

COOLANT BYPASS

COOLANT PUMP

SUPPLY LINE

B

ENGINE COOLANT INLET

OPEN-STAT COOLANT FLOW (PARTIAL-BLOCKING-TYPE STAT)

Fig. 11-6. Self-purging system. (Courtesy International Harvester.)

temperatures than the radiator would allow. On the other hand, coolant in the radiator is at near ambient temperature, which means that it can freeze even while the engine is running. Antifreeze is mandatory.

Marine applications may employ a raw water system as illustrated in Fig. 11-8. The scoop (shown at *A*) should be of standard marine design, with generous-sized fittings, and barred to give some protection to the raw water pump and heat exchanger. In addition you may wish to install a strainer (*D*). The seacock (*B*) should be the gate valve type that opens

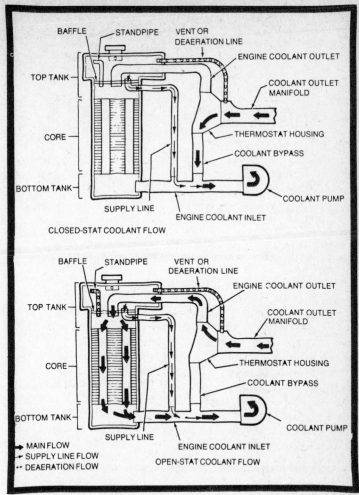

CLOSED-STAT COOLANT FLOW

OPEN-STAT COOLANT FLOW

→ MAIN FLOW
→ SUPPLY LINE FLOW
⋯ DEAERATION FLOW

Fig. 11-7. Full deaeration system. (Courtesy International Harvester.)

fully and should have a minimum orifice size of 1 in. NPT (National Pipe Thread) for engines of 200 hp and under. If hosing is used it should be reinforced to withstand the suction of the pump.

Figure 11-9 shows the layout for a fresh water system, which is, in most ways, preferable to using sea water directly as a coolant. The heat exchanger is by far the most critical component and is generally fitted with removable caps to facilitate cleaning. Periodically rod the tubes with a wooden dowel. Many exchangers have an additional refinement in the shape of a zinc "pencil." The zinc is sacrificial and is attacked

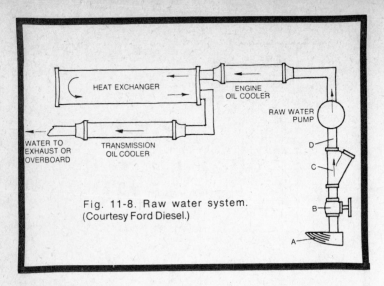

Fig. 11-8. Raw water system.
(Courtesy Ford Diesel.)

before the metal of the exchanger. Replace the "pencil" as
needed.

If you opt for a dry manifold you may—depending upon
engine size, hull shape, and boat use—be able to dispense with
the raw water pump. Consult with your dealer.

Perhaps the best test of a closed system is to bleed a
known amount of water relative to the total system capacity
and observe the action in the header tank under normal loads.
The amount of coolant the system can lose without loss of flow
or entrapment of air is known as the *drawdown rating*. Figure
11-10 shows acceptable drawdown capacities for systems of up
to 360 quarts capacity.

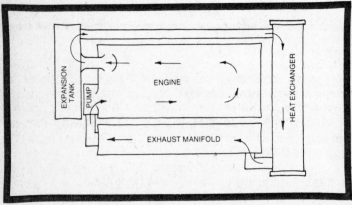

Fig. 11-9. Fresh water system. (Courtesy Lehman Ford Diesel.)

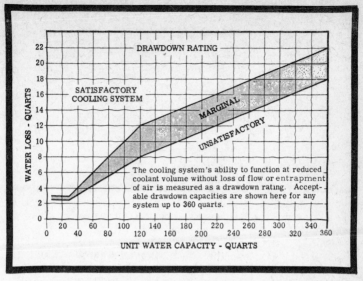

Fig. 11-10. Drawdown capacities. (Courtesy International Harvester.)

## Radiators

Radiator core designs should be tailored to the application. The canted-tube Z-core is one of the most popular since the lay of the tubes generates air turbulence which "scrubs" the heat away. The straightforward wood core is used in dusty areas—in logging, demolition, rock-quarrying, and underbrush-clearing operations—and is characterized by wide air passages and tubes stacked in parallel. Most cores are made of copper, although aluminum has been tried, and steel is sometimes used in desert and other abrasive environments.

The radiator fins should be blown free of debris as needed. Lift trucks used to handle cotton or jute must have their radiator passages cleared every 30 min. or so. Over-the-highway vehicles can go for years without this service.

Coolant level is another critical factor. It should come within an inch of the bottom of the filter neck or to the mark inscribed on the fillter or overflow tank. Sometimes radiators will fool you: there may be an air bubble trapped below the baffle (Fig. 11-11). Double-check with the engine running and the thermostat open.

The radiator should be reverse-flushed as needed (Fig. 11-12). Drain the system and disconnect both hoses. Connect the flushing gun to the lower hose and to an air source of 100 psi. Turn on the water. When the radiator is full, inject air in

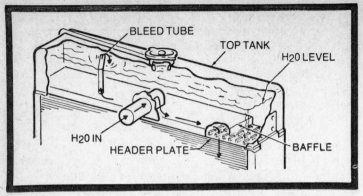

Fig. 11-11. Top-tank construction. The baffle helps to separate gases and coolant by spreading the outlet flow from the engine over the core with very little turbulence. (Courtesy International Harvester.)

short bursts to clear the tubes. Between bursts allow the radiator to fill. Continue until the water flows freely or until the radiator passes a flow test. This test is a measure of the amount of water which falls through the radiator in one minute and, of course, depends upon the manufacturer's specifications. Radiators can be cleaned with commercial preparations (some of which are not compatible with antifreeze) or dismantled and boiled. In extreme cases the header tank must be unsoldered and the tubes rodded.

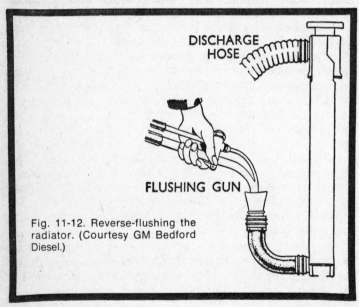

Fig. 11-12. Reverse-flushing the radiator. (Courtesy GM Bedford Diesel.)

286

Leaks are usually quite obvious and tend to appear around the joints of the header tank and tubes. Elusive leaks can be pinpointed by placing the radiator in a tub of water and blocking the outlet and applying no more than 15 psi of air to the inlet. Less test pressure is advisable if the radiator is old and fragile.

Most shops prefer to farm out radiator work, having had unfortunate experiences with their own radiator work. But unless the leak is buried within the tube columns or unless other repairs are needed, such as replacement of fins or removal of the header tank, the average mechanic can cope with radiators. The trick is in the soldering.

Use 60/40 straight bar or acid-core solder. Mix a few ounces of flux. The traditional formula is muriatic (dilute hydrochloric) acid and zinc powder. Clean the joint and reclean it. With a pencil torch, heat the area to be mended. Do not employ too much heat and be careful not to heat adjacent tubes or joints. Apply the solder. If it bubbles and skates, the surface is still not clean. It should sink into the copper (or steel), leaving a mirror-like glaze on the surface which becomes dull as it hardens. Allow the solder to cool at its own rate and test the radiator in the water tank.

**Pressure Caps**. Standard practice is to pressurize the cooling system at a few psi above atmospheric pressure. The additional pressure raises the boiling point of the coolant and allows hotter operation. Internal combustion engines operate best at some temperature approaching 200°F. Another advantage of pressurization is that it delays boiling at high altitudes.

Figure 11-13 illustrates a typical pressure cap. Note that it is somewhat complex, with a seal on the radiator filler tube flange and two valves. The main valve is the pressure relief and opens at 3—7 psi, depending upon application. The vacuum valve is located inside the spring and equalizes pressures after shutdown.

A failed pressure cap can cause loss of coolant from boiling and, consequently, overheating of the engine. Caps should be inspected visually and, if the occasion demands it, tested. AC and other suppliers furnish pressure testers which can be used to test the cap and cooling system.

Although everyone should know this by now, I emphasize that pressure caps are designed to be opened in two stages. The first stage bleeds down the pressure and directs the hot vapor down, away from the mechanic's hand. The second stage releases the cap. Open one too soon and you can get burned.

287

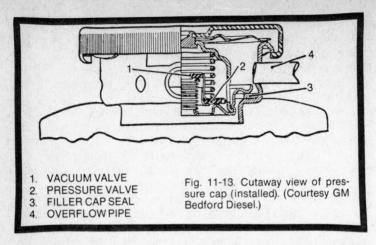

1. VACUUM VALVE
2. PRESSURE VALVE
3. FILLER CAP SEAL
4. OVERFLOW PIPE

Fig. 11-13. Cutaway view of pressure cap (installed). (Courtesy GM Bedford Diesel.)

**Radiator Hoses.** Hoses are a necessary evil since, in most applications, the engine is free to rock on its mounts, while the radiator is fixed. Eventually hoses will have to be replaced since they fail by the combined action of flexzing and heat. Inspect the outside covering for cracks and bulges, especially at the mounting points. Squeeze the hoses to determine if they are still resilient. It is not unusual for the lower hose to collapse from pump sunction as the engine runs. Heater hoses should be valved at the engine, and not at the heater. Otherwise stagnant water collects in them during the summer months.

Many patented hose clamps are in use, but most mechanics prefer the stainless steel type with a worm gear that engages serrations on the ribbon. Spring clamps are generally found on late-model equipment and are adequate.

### Thermostats

The thermostat is located in a removable housing at the fan end of the engine. Single or double thermostats may be fitted with opening temperatures of from about 150°F to 205°F. The "colder" thermostats were intended to be used with ethyl or methyl alcohol antifreeze. Alcohol boils at 178°F.

The illustration in Fig. 11-14 shows a pill or pellet type, opened and closed. When closed, a small amount of coolant flows through the seepage port to the radiator, but most is recirculated through the bypass valve (No. 8 in the drawing) to the pump and back through the engine. As the engine warms, the wax in the pellet goes into the liquid state and expands. It forces out the pin attached to the frame (No. 4). This action pries open the valve (2) against the spring (7) and clears the

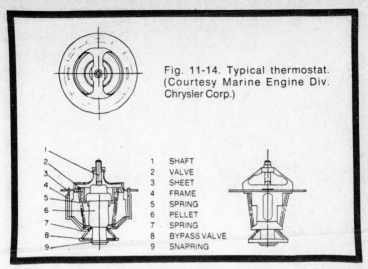

Fig. 11-14. Typical thermostat. (Courtesy Marine Engine Div. Chrysler Corp.)

| | |
|---|---|
| 1 | SHAFT |
| 2 | VALVE |
| 3 | SHEET |
| 4 | FRAME |
| 5 | SPRING |
| 6 | PELLET |
| 7 | SPRING |
| 8 | BYPASS VALVE |
| 9 | SNAPRING |

coolant passage to the radiator. The action is progressive. In the partially open condition the coolant flow splits between the bypass valve and radiator. At full open all of the coolant is directed to the radiator.

Check the thermostat for signs of physical damage. Frames tend to break, and the metal bellows used on many (rather than telescoping tubes shown in Fig. 11-14) can develop cracks at the seams.

To check the action, heat the thermostat in water, supporting it as shown in Fig. 11-15. The temperature rating, usually found stamped on the frame, refers to the temperature at which the thermostat just cracks open. The full-open

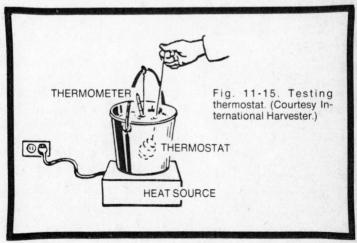

THERMOMETER

Fig. 11-15. Testing thermostat. (Courtesy International Harvester.)

THERMOSTAT

HEAT SOURCE

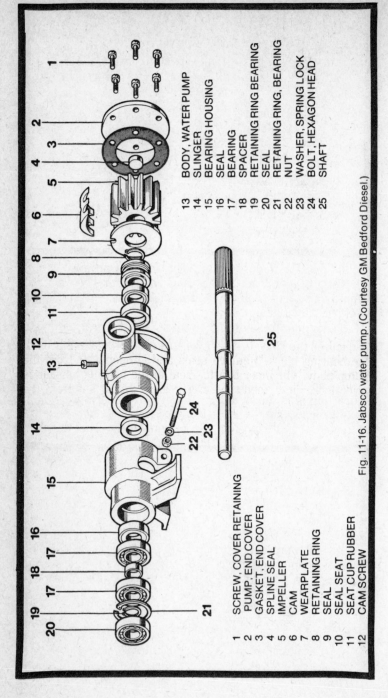

1 SCREW, COVER RETAINING
2 PUMP, END COVER
3 GASKET, END COVER
4 SPLINE SEAL
5 IMPELLER
6 CAM
7 WEARPLATE
8 RETAINING RING
9 SEAL
10 SEAL SEAT
11 SEAT CUP RUBBER
12 CAM SCREW

13 BODY, WATER PUMP
14 SLINGER
15 BEARING HOUSING
16 SEAL
17 BEARING
18 SPACER
19 RETAINING RING BEARING
20 SEAL
21 RETAINING RING, BEARING
22 NUT
23 WASHER, SPRING LOCK
24 BOLT, HEXAGON HEAD
25 SHAFT

Fig. 11-16. Jabsco water pump. (Courtesy GM Bedford Diesel.)

temperature can be found in the engine builder's specs and should be tested. A partially open thermostat will run you around in circles. High-temperature stats can be tested in a solution of ethylene glycol and water. A 50/50 mix has a boiling point of 227°F.

## Water Pumps

Raw water pumps are rugged devices designed to be mounted at some point remote from the engine. The Jabsco unit in Fig. 11-16 is one of the most popular. It has prelubricated bearings and a symmetrical impeller, which means it can be driven in either direction. Intake and discharge ports are at the top of the pump body and must be swapped to match changes in rotation. The pump is lubricated by water and should not be run dry for longer than it takes to prime. This model has no drain cock. In freezing temperatures the end cover screws are loosened to drain.

To disassemble, remove the cover retaining screws, end cover (No. 2), and impeller (5). If the pump has any time on it, the impeller will be stuck. Pull it off of the shaft with pliers on two blades. Remove the cam locking screw (13) and the cam (6). The wear plate (7) is now accessible. Replace it as a routine precautionary matter. Support the pump body in a vise and loosen the pinch bolt (24) which secures the pump body to the bearing block. Remove seals and slinger (14) from the pump body. The inner bearing seal in the bearing housing is pried loose first, then the outer seal is driven out from the impeller side (after the shaft and bearings have been pressed out). The last part to remove is the spring clip (19) which secures the bearings to the shaft.

Replace all seals and any other parts which show wear or corrosion damage. Assemble in the reverse order of disassembly. The cam screw (13) is as critical as any other single fastener in the system. Should it vibrate loose, the cam will drop down into the casting and destroy or damage the pump. Secure it with Loctite or the equivalent.

The engine pump (Fig. 11-17) is integral to the cooling system and is present in all applications. Some designs employ the *thermosiphon* principle—the tendency of hot water to rise and displace cold water—but no modern engine depends entirely on this. The pump has been refined to stark simplicity over the years and consists of a pump body, centrifugal impeller, shaft, bearing, and seal pack. No routine maintenance is required (other than keeping the system clean). The pump should be opened only if one or more of the following symptoms is present:

- Perceptible friction as the shaft is turned by hand.
- Leaks around the shaft bearing.
- Leaks from the vent hole on the pump body. Some moisture is normal and represents seepage past the seal. But droplets or a continuous stream discharged here point to seal and possibly bearing failure.

To disassemble, remove the fan and detach the pump body from the engine block. Fan hub, impeller, and bearing removal require a 5-ton arbor press. Rebuild kits are available with shaft, impeller, seals, and gaskets (Fig. 11-18).

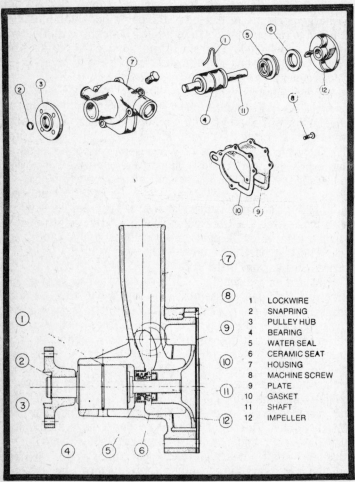

| | | |
|---|---|---|
| | 1 | LOCKWIRE |
| | 2 | SNAPRING |
| | 3 | PULLEY HUB |
| | 4 | BEARING |
| | 5 | WATER SEAL |
| | 6 | CERAMIC SEAT |
| | 7 | HOUSING |
| | 8 | MACHINE SCREW |
| | 9 | PLATE |
| | 10 | GASKET |
| | 11 | SHAFT |
| | 12 | IMPELLER |

Fig. 11-17. Pump in two views. (Courtesy Marine Engine Div., Chrysler Corp.)

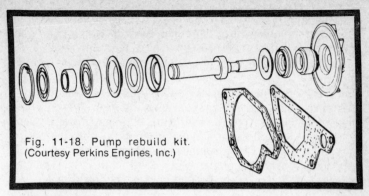

Fig. 11-18. Pump rebuild kit. (Courtesy Perkins Engines, Inc.)

## Engine Block

The cutaway drawing in Fig. 11-19 illustrates the typical flow pattern through the block and head. Coolant flow is concentrated around the injectors, valve seats, swirl chambers, and valve guides. Although not shown here, a number of engines employ baffles (sometimes called directors) placed in the head casting to direct water over these

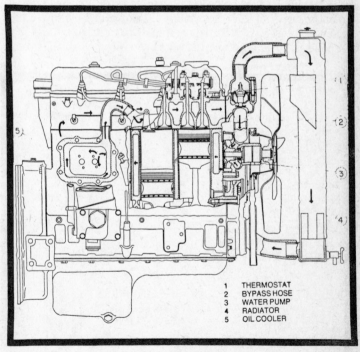

| 1 | THERMOSTAT |
| 2 | BYPASS HOSE |
| 3 | WATER PUMP |
| 4 | RADIATOR |
| 5 | OIL COOLER |

Fig. 11-19. Coolant circulation. (Courtesy Marine Engine Div., Chrysler Corp.)

critical parts. The modern tendency is to keep the water inlet and outlet temperatures within a narrow range to reduce stress and to allow high coolant velocity. Low-velocity systems tended to leave a layer of steam clinging to the hot side of the jacket.

The block should be inspected for telltale rust stains at the gasketed joints, and especially around the freeze plugs. These plugs deteriorate from the inside out and should be replaced every few years. If expansion plugs are used, coat the mating surface with 3M weather adhesive cement or a comparable product and drive the plug into place with a mallet. Threaded plugs give better security and should be used whenever possible.

The major difficulty with the water jacket is rust and scale formation. According to General Motors, $1/16$ in. of scale adhering to a 1 in. cast-iron wall is the equivalent in terms of insulation to $4\frac{1}{4}$ in. of cast iron. Loose scale can be removed with a flushing gun as shown in Fig. 11-20. Remove the thermostat and plug the heater hoses. Operate the gun at 100 psi, allowing the jacket to fill with water between air bursts.

Chemicals may be used to dissolve scale and sludge, but do not expect miracles, especially if the water in your locality is hard. The really effective chemicals are corrosive and may damage the radiator and heater core. All such preparations

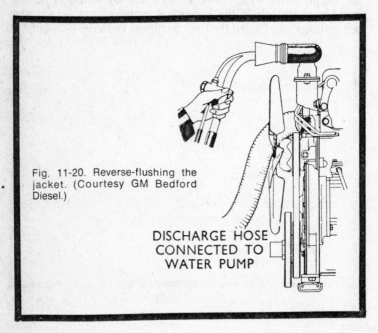

Fig. 11-20. Reverse-flushing the jacket. (Courtesy GM Bedford Diesel.)

DISCHARGE HOSE
CONNECTED TO
WATER PUMP

should be neutralized by following the instructions on the container to the letter. The best way to clean a block is to have it boiled. Automotive machinists have the necessary equipment. The parts to be boiled must be stripped of all nonferrous metal—including the camshaft bearings.

## Coolant

Ideally no diesel engine should operate without a water filter and conditioner. If the engine is already in service, drain and flush the coolant. It may be necessary to change the filter element at close intervals during the first few hundred hours of operation if the jacket has rusted or silted. Each time the filter is changed, buff the metal element on a wire brush to expose new metal to the coolant. Replace the element as needed. Filters which employ sacrificial elements must be securely grounded to the engine block or mounting-frame members.

Lenrock nonchromate filters give good year-round protection and can tolerate hard water better than most. (avoid filters with magnesium parts). But, even with the best filter and conditioner, common sense tells you that the water supply must be carefully evaluated. In general, water should have a total hardness of under 170 ppm (parts per million), and fewer than 100 ppm of sulfates and 40 ppm of chlorides. Dissolved solids should be kept below 340 ppm.

If these contaminants are present, distill or demineralize the supply. The water may be softened by addition of the suitable chemicals if the individual contaminant levels are below those listed and if the total hardness is above 170 ppm.

Inhibitor compounds are available from several manufacturers. They are used singly in bulk form; premixed with antifreeze; or in concert with a filter—conditioner. Obtain detailed information from your dealer before using any of these products, since they may not be compatible.

For example, sodium chromate and potassium dichromate are not compatible with ethylene glycol antifreeze. On contact with permanent antifreeze these compounds produce a green slime which will cut the rate of heat transfer enough to cause overheating. It can be removed by flushing, followed by a dose of descaler.

Nonchromate inhibitors—a class which includes nitrates, nitrides, and borates—is recommended because of the convenience. These can be used with permanent-type antifreeze solutions or with water. Soluble oil is now obsolete as a corrosion inhibitor. At concentrations of more than 1% by volume, heat transfer becomes problematic. GM engineers have found that a 2½% solution increases fire deck temperature by as much as 15%.

Alcohol should not be used as antifreeze except in an emergency. It boils at lower temperature than operating temperature and must constantly be replenished.

Ethylene glycol is permanent and generally contains rust inhibitors. These inhibitors should be replenished at 500 hr with a nonchromate inhibitor such as Nalcool 2000. Do not use antifreeze with sealant.

In general, mix antifreeze with softened water in a 50/50 proportion. This will give freeze protection to −40°F and will raise the boiling point to approximately 227°F. At between 62% and 75% the engine will be protected to almost −70°F, with a commensurate rise in the boiling point. But greater concentrations will lower the freezing point. Under no condition should the gylcol mix be richer than 67/40.

The major drawback associated with ethylene glycol is its reaction to oil. Should a leak develop at the head gasket or oil cooler, shut down immediately. Assuming that the engine is in running condition, you may be able to save a complete teardown by flushing with Butyl Cellosolve. This procedure was developed for Detroit Diesel 2-cycles and should not necessarily be employed in other engines, which may have higher bearing loadings. Query your dealer or factory rep. The procedure is as follows:

1. Drain the sump.
2. Change the lube oil filter element and clean the housing.
3. Mix two parts Butyl Cellosolve with one part SAE 10 oil.
4. Run the engine at 1000−1200 rpm for 1 hr. Keep a sharp eye on the oil pressure gage since lubrication is only marginal.
5. Drain, allowing all of the oil to escape, and refill with SAE 10 oil.
6. Run for a quarter-hour at the same rpm as before.
7. Drain, replace the filter, and fill the sump up to the mark with standard-grade oil.
8. Run for 30 min, shut down, and restart. If the starter drags, you can be fairly sure that there is still ethylene glycol in the system. Flush again, repeating each step of the process.

Glycol ether (methoxypropanol) has been suggested as an alternative to ethylene glycol. However, it is reported to attack rubber parts, such as the Viton rubber O-rings in Cummins engines, as well as GM head gaskets.

## Fans

Most diesel engines have simple, single-speed fans, although in vehicles they may have magnetic- or viscous-drive fans. The reason for this is to save fuel. At more than 35 mph or so, the fan is not needed and absorbs several horsepower which could be better used to turn the crankshaft. In general these fans are not repairable, although magnetic coils are sometimes available.

Fixed-speed fans can, depending upon the design, be installed backwards. Note the lay of the blades before removal. Most applications use the fan in suction, although on industrial trucks and the like the fan often blows through the radiator.

Cooling efficiency can be increased by modifications to the fan. These modifications should not be undertaken in lieu of repairs to the radiator or other components.

In order of cost progression these modifications are:

1. Move the fan closer to the radiator by means of spacers on the hub. Leave at least ½ in. clearance between the radiator and fan.
2. Employ a venturi-type fan ring. Many industrial engines merely have a box around the fan or a shield at the header tank.
3. Change the fan to a more efficient version. Some consultation with your supplier will be necessary to determine the optimum pitch and number of blades.
4. Change the drive ratio to speed the fan. Contact your dealer before you make this modification since the fan is mounted on the water pump shaft.

Drive belts should be tightened routinely to specifications. Some give is necessary to protect the bearings and prolong belt life. Examine the belt for oil damage, wear on the flanks, and heat cracks on its internal surface. Any of these faults means that the belt should be replaced. Paired belts are replaced together. Otherwise, sag will develop in the used belt, causing it to slip and wear the pulley flange. International Harvester engineers have measured wear rates under these conditions and report that the ratio is 1:5, with the belt taking the brunt of it. The two drawings in Fig. 11-21 compare a new belt with one that should be discarded. Note that the worn belt has fallen into the bottom of the groove. Any torque it transmits will be at this point, and not on the groove sides.

## COOLING-SYSTEM ACCESSORIES

Turbocharged engines may be fitted with an intercooler between the turbine outlet and engine intake. The purpose of the intercooler is to reduce charge temperature and thus increase its density for more power. Maintenance is straightforward, calling for no special techniques. The tubes

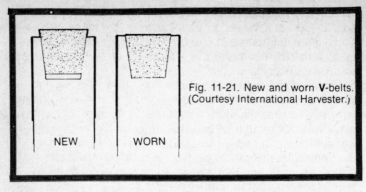

Fig. 11-21. New and worn **V**-belts. (Courtesy International Harvester.)

NEW    WORN

are accessible after disassembly and may be cleaned with a brush. If leaks are suspected the unit may be checked by immersing it in water and slightly pressurizing the tube assembly.

Figure 11-22 shows a combined heat exchanger and oil cooler of the type widely used in marine service. Three elements are combined into a single assembly: a header tank for the fresh water system, a fresh water tube cluster, and an oil tube cluster.

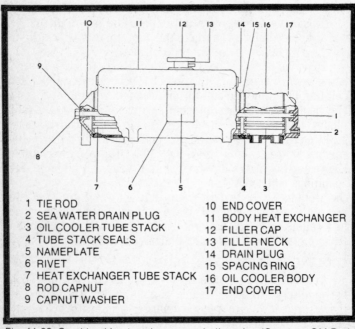

| 1 | TIE ROD | 10 | END COVER |
| 2 | SEA WATER DRAIN PLUG | 11 | BODY HEAT EXCHANGER |
| 3 | OIL COOLER TUBE STACK | 12 | FILLER CAP |
| 4 | TUBE STACK SEALS | 13 | FILLER NECK |
| 5 | NAMEPLATE | 14 | DRAIN PLUG |
| 6 | RIVET | 15 | SPACING RING |
| 7 | HEAT EXCHANGER TUBE STACK | 16 | OIL COOLER BODY |
| 8 | ROD CAPNUT | 17 | END COVER |
| 9 | CAPNUT WASHER | | |

Fig. 11-22. Combined heat exchanger and oil cooler. (Courtesy GM Bedford Diesel.)

To disassemble, remove the fresh water lines and the rod capnut (No. 8). The end covers are now free. Support the oil cooler (16) and the spacing ring (15) since the throughbolt is all that holds these parts to the main casting. Clean the tube stack chemically or with a ⅛ in. brass rod inserted against the direction of water flow. The rod should be dulled on the end, without sharp edges, and must be slowly worked into the tubes. Pressure-test the exchange circuits at 30 psi and the oil cooler at 100 psi. Plugs will have to be fabricated to hold the pressure. The cooler section should be immersed in simmering water to open any holes prior to the test.

# Index

# Index

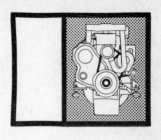